T0136286

Essentials of CLOUD COMPUTING

Essentials of CLOUD COMPUTING

K. Chandrasekaran

CRC Press
Taylor & Francis Group
Boca Raton London New York

CRC Press is an imprint of the
Taylor & Francis Group, an **informa** business

A CHAPMAN & HALL BOOK

CRC Press
Taylor & Francis Group
6000 Broken Sound Parkway NW, Suite 300
Boca Raton, FL 33487-2742

Printed on acid-free paper
Version Date: 20141014

International Standard Book Number-13: 978-1-4822-0543-5 (Hardback)

Visit the Taylor & Francis Web site at
http://www.taylorandfrancis.com

and the CRC Press Web site at
http://www.crcpress.com

Contents

Foreword

Cloud computing is sprawling the IT landscape. Driven by several converging and complementary factors, cloud computing is advancing as a viable IT service delivery model at an incredible pace. It has caused a paradigm shift in how we deliver, use, and harness the variety of IT services it offers. It also offers several benefits compared to traditional on-premise computing models, including reduced costs and increased agility and flexibility. Its transformational potential is huge and impressive, and consequently cloud computing is being adopted by individual users, businesses, educational institutions, governments, and community organizations. It helps close the digital (information) divide. It might even help save our planet by providing an overall greener computing environment.

Hence, corporations are eagerly investing in promising cloud computing technologies and services not only in developed economies but also increasingly in emerging economies—including India, China, Taiwan, the Philippines, and South Africa—to address a region's specific needs. Cloud computing is receiving considerable interest among several stakeholders—businesses, the IT industry, application developers, IT administrators and managers, researchers, and students who aspire to be successful IT professionals.

To successfully embrace this new computing paradigm, however, they need to acquire new cloud computing knowledge and skills. In answer to this, universities have begun to offer new courses on cloud computing. Though there are several books on cloud computing—from basic books intended for general readers to advanced compendium for researchers—there are few books that comprehensively cover a range of cloud computing topics and are particularly intended as an entry-level textbook for university students. This book, *Essentials of Cloud Computing*, fills this void and is a timely and valuable addition by Professor K. Chandrasekaran, a well-recognized academic and researcher in cloud computing.

The book, beginning with a brief overview on different computing paradigms and potentials of those paradigms, outlines the fundamentals of cloud computing. Then, it deals with cloud services types, cloud deployment models, technologies supporting and driving the cloud, software process models and programming models for cloud, and development of software application that runs the cloud. It also gives an overview of services available from major cloud providers, highlights currently available open source software and tools for cloud deployment, and discusses security concerns and issues in cloud computing. Finally, it outlines advances in cloud computing such as mobile cloud and green cloud. The book's presentation style supports ease of reading and comprehension. Further, each chapter is supplemented with review questions that help the readers to check their understanding of topics and issues explored in the chapter.

Cloud computing is here to stay, and its adoption will be widespread. It will transform not only the IT industry but also every sector of society. A wide range of people—application developers, enterprise IT architects and administrators, and future IT professionals and managers—will need to learn about cloud computing and how it can be deployed for a variety of applications. This concise and comprehensive book will help readers understand several key aspects of cloud computing—technologies, models, cloud services currently available, applications that are better suited for cloud, and more. It will also help them examine the issues and challenges and develop and deploy applications in clouds.

I believe you will find the book informative, concise, comprehensive, and helpful to gain cloud knowledge.

San Murugesan
Adjunct Professor
University of Western Sydney
Richmond, New South Wales, Australia
Editor in Chief
IEEE IT Professional
Editor
IEEE Computer
Director
BRITE Professional Services
Australia

Preface

Cloud computing is one of the most popular technologies that has become an integral part of the computing world nowadays. The usage and popularity of cloud computing is increasing every day and is expected to increase further. Many frequent Internet users are heavily dependent on cloud-based applications for their day-to-day activities in both professional and personal life. Cloud computing has emerged as a technology to realize the utility model of computing while using Internet for accessing applications.

The past decades have witnessed the success of centralized computing infrastructures in many application domains. Then, the emergence of the Internet brought numerous users of remote applications based on the technologies of distributed computing. Research in distributed computing gave birth to the development of grid computing. Though grid is based on distributed computing, the conceptual basis for grid is somewhat different. Computing with grid enabled researchers to do computationally intensive tasks by using limited infrastructure that was available with them and with the support of high processing power that could be provided by any third party, and thus allowing the researchers to use grid computing, which was one of the first attempts to provide computing resources to users on payment basis. This technology indeed became popular and is being used even now. An associated problem with grid technology was that it could only be used by a certain group of people and it was not open to the public. Cloud computing in simple terms is further extension and variation of grid computing, in which a market-oriented aspect is added. Though there are several other important technical differences, this is one of the major differences between grid and cloud. Thus came cloud computing, which is now being used as a public utility computing software and is accessible by almost every person through the Internet. Apart from this, there are several other properties that make cloud popular and unique. In cloud, the resources are metered, and a user pays according to the usage. Cloud can also support a continuously varying user demands without affecting the performance, and it is always available for use without any restrictions. The users can access cloud from any device, thus reaching a wider range of people.

There are several applications of cloud computing already being witnessed and experienced. As cloud is elastic, it can be used in places where varying load is one of the main characteristics. It can also be used where on-demand access is required. Similarly, because of its property of multitenancy, it can be used in places where several applications are to be operated. Cloud computing can also be used for data-intensive applications for data analytics and several data-related tasks.

As this is considered a promising technology, several companies such as Google, Microsoft, Amazon, HP, and IBM have invested their time and other resources for further development of cloud computing–related technologies. In return, the companies make profit as cloud applications become more popular and easier to use.

The main objective of this book is to present the readers with the introductory details, technologies, applications development, security, and some advanced topics in cloud computing. It is expected that the book will serve as a reference for a larger audience base, including students of undergraduate and postgraduate programs, practitioners, developers, and new researchers.

This book will be a timely contribution to cloud computing, a field that is gaining momentum in all dimensions such as academia, research, and business. As cloud computing is recognized as one of the top five emerging technologies that will have a major impact on the quality of science and society over the next 20 years, its knowledge will help position our readers at the forefront of the field.

This book discusses in detail the essentials of cloud computing in a way suitable for beginners as well as practitioners who are in need to know or learn more about cloud computing. It can also be used as a handbook for cloud. It contains 14 chapters that follow a standard format for easy and useful reading: Learning Objectives, Preamble, Introduction, and details related to the chapter title with several subsections, supported by a suitable number of diagrams, tables, figures, etc., followed by Summary, Review Points, Review Questions, and References for further reading.

To start with, Chapter 1 aims to give a brief description about available paradigms of computing. This provides the required basic knowledge about computing paradigms to start with cloud technology. It includes several computing paradigms such as high-performance computing, cluster computing, grid computing, and distributed computing.

Chapter 2 gives a basic introduction and discusses the fundamental concepts of cloud. The topics include cloud computing definition, the need for cloud, cloud principles, cloud applications, and several other topics.

Chapter 3 gives an introduction to cloud computing technologies. This includes the basic concepts in cloud such as cloud architecture, cloud anatomy, network connectivity in cloud, cloud management, applications in cloud, and migrating applications to cloud.

Chapter 4 discusses in detail the deployment models such as private, public, community, and hybrid. Their applications, use, and design are also discussed, thereby giving a clear picture and facilitating a proper choice of deployment models.

Chapter 5 discusses in detail the cloud service models such as Software as a Service (SaaS), Platform as a Service (PaaS), and Infrastructure as a Service (IaaS) with several other service models that have emerged recently.

This chapter gives an idea on the properties, architecture, and application of these cloud service models.

Chapter 6 discusses the technological drivers for cloud. The topics covered in these chapters are service-oriented architecture and cloud, virtualization, multicore technology, software models for cloud, pervasive computing, and several other related concepts. This chapter gives an elaborate view on how these technological drivers are related to cloud and promote it further in the context of application development and research.

Chapter 7 gives a detailed description about virtualization. Virtualization is considered to be the basis of cloud computing. Here, opportunities and approaches related to virtualization are discussed. Hypervisors are discussed in detail. This chapter also gives a description on how virtualization is used in cloud computing.

Chapter 8 discusses the programming models that are available for cloud. Here, both existing programming models useful to migrate to cloud and new programming models specific to cloud are discussed in detail.

Chapter 9 describes cloud from a software development perspective, the different perspectives of SaaS development and its challenges, and cloud-aware software development in PaaS.

Chapter 10 deals with the networking aspects in the cloud computing environment. This chapter also presents an overview and issues related to the data center environment. Transport layer issues and Transmission Control Protocol enhancements in data center networks are also discussed.

Chapter 11 gives a brief description of major service providers known in the cloud arena and discusses in detail about the services they offer.

Chapter 12 is especially for open-source users. This chapter gives a list and description of several open-source support and tools available for cloud computing. These are divided according to the service models, that is, SaaS, PaaS, and IaaS. There is also a separate note on open-source tools for research, which describes the tools that can be worked upon in from a research-oriented perspective. It also has an additional note on distributed computing tools that are used for managing distributed systems.

Chapter 13 discusses the security issues in cloud, an important issue in cloud. It discusses about security aspects in general, platform-related security, audit, and compliance in cloud.

The final chapter, Chapter 14, discusses advanced concepts in cloud, such as intercloud, cloud management, mobile cloud, media cloud, cloud governance, green cloud, cloud analytics, and several other allied topics.

The contents of this book reflect the author's lectures on this topic. The author wishes to acknowledge the following for their valuable time and contributions in developing, improving, and formatting the chapters: Mohit P. Tahiliani, Marimuthu C., Raghavan S., Manoj V. Thomas, Rohit P. Tahiliani, Alaka A., Usha D., Anithakumari S., and Christina Terese Joseph.

Finally, the author would like to thank Aastha Sharma, commissioning editor at CRC Press, for her constant communication and follow-up and support throughout the process of getting the book into print.

Readers are requested to visit the website http://www.cloudrose.org/ for further updates and e-mail interactions with the author.

K. Chandrasekaran

1

Computing Paradigms

Learning Objectives

The objectives of this chapter are to

- Give a brief description of major of computing
- Examine at the potential of these paradigms

Preamble

The term paradigm conveys that there is a set of practices to be followed to accomplish a task. In the domain of computing, there are many different standard practices being followed based on inventions and technological advancements. In this chapter, we look into the various computing paradigms: namely high performance computing, cluster computing, grid computing, cloud computing, bio-computing, mobile computing, quantum computing, optical computing, nanocomputing, and network computing. As computing systems become faster and more capable, it is required to note the features of modern computing in order to relate ourselves to the title of this book on cloud computing, and therefore it becomes essential to know little on various computing paradigms.

1.1 High-Performance Computing

In high-performance computing systems, a pool of processors (processor machines or central processing units [CPUs]) connected (networked) with other resources like memory, storage, and input and output devices, and the deployed software is enabled to run in the entire system of connected components.

The processor machines can be of homogeneous or heterogeneous type. The legacy meaning of high-performance computing (HPC) is the supercomputers; however, it is not true in present-day computing scenarios. Therefore, HPC can also be attributed to mean the other computing paradigms that are discussed in the forthcoming sections, as it is a common name for all these computing systems.

Thus, examples of HPC include a small cluster of desktop computers or personal computers (PCs) to the fastest supercomputers. HPC systems are normally found in those applications where it is required to use or solve scientific problems. Most of the time, the challenge in working with these kinds of problems is to perform suitable simulation study, and this can be accomplished by HPC without any difficulty. Scientific examples such as protein folding in molecular biology and studies on developing models and applications based on nuclear fusion are worth noting as potential applications for HPC.

1.2 Parallel Computing

Parallel computing is also one of the facets of HPC. Here, a set of processors work cooperatively to solve a computational problem. These processor machines or CPUs are mostly of homogeneous type. Therefore, this definition is *the same* as that of HPC and is broad enough to include supercomputers that have hundreds or thousands of processors interconnected with other resources. One can distinguish between *conventional* (also known as serial or sequential or Von Neumann) computers and parallel computers in the way the applications are executed.

In serial or sequential computers, the following apply:

- It runs on a single computer/processor machine having a single CPU.
- A problem is broken down into a discrete series of instructions.
- Instructions are executed one after another.

In parallel computing, since there is simultaneous use of multiple processor machines, the following apply:

- It is run using multiple processors (multiple CPUs).
- A problem is broken down into discrete parts that can be solved concurrently.
- Each part is further broken down into a series of instructions.

- Instructions from each part are executed simultaneously on different processors.
- An overall control/coordination mechanism is employed.

1.3 Distributed Computing

Distributed computing is also a computing system that consists of multiple computers or processor machines connected through a network, which can be homogeneous or heterogeneous, but run as a single system. The connectivity can be such that the CPUs in a distributed system can be physically close together and connected by a local network, or they can be geographically distant and connected by a wide area network. The heterogeneity in a distributed system supports any number of possible configurations in the processor machines, such as mainframes, PCs, workstations, and minicomputers. The goal of distributed computing is to make such a network work as a single computer.

Distributed computing systems are advantageous over centralized systems, because there is a support for the following characteristic features:

1. Scalability: It is the ability of the system to be easily expanded by adding more machines as needed, and vice versa, without affecting the existing setup.
2. Redundancy or replication: Here, several machines can provide the same services, so that even if one is unavailable (or failed), work does not stop because other similar computing supports will be available.

1.4 Cluster Computing

A cluster computing system consists of a set of the same or similar type of processor machines connected using a dedicated network infrastructure. All processor machines share resources such as a common home directory and have a software such as a message passing interface (MPI) implementation installed to allow programs to be run across all nodes simultaneously. This is also a kind of HPC category. The individual computers in a cluster can be referred to as *nodes*. The reason to realize a cluster as HPC is due to the fact that the individual nodes can work together to solve a problem larger than any computer can easily solve. And, the nodes need to communicate with one another in order to work cooperatively and meaningfully together to solve the problem in hand.

If we have processor machines of heterogeneous types in a cluster, this kind of clusters become a subtype and still mostly are in the experimental or research stage.

1.5 Grid Computing

The computing resources in most of the organizations are underutilized but are necessary for certain operations. The idea of grid computing is to make use of such nonutilized computing power by the needy organizations, and thereby the return on investment (ROI) on computing investments can be increased.

Thus, grid computing is a network of computing or processor machines managed with a kind of software such as middleware, in order to access and use the resources remotely. The managing activity of grid resources through the middleware is called *grid services*. Grid services provide access control, security, access to data including digital libraries and databases, and access to large-scale interactive and long-term storage facilities.

TABLE 1.1

Electrical Power Grid and Grid Computing

Electrical Power Grid	Grid Computing
Never worry about where the electricity that we are using comes from; that is, whether it is from coal in Australia, from wind power in the United States, or from a nuclear plant in France, one can simply plug the electrical appliance into the wall-mounted socket and it will get the electrical power that we need to operate the appliance.	*Never worry* about where the computer power that we are using comes from; that is, whether it is from a supercomputer in Germany, a computer farm in India, or a laptop in New Zealand, one can simply plug in the computer and the Internet and it will get the application execution done.
The infrastructure that makes this possible is called *the power grid*. It links together many different kinds of power plants with our home, through transmission stations, power stations, transformers, power lines, and so forth.	*The infrastructure* that makes this possible is called *the computing grid*. It links together computing resources, such as PCs, workstations, servers, and storage elements, and provides the mechanism needed to access them via the Internet.
The power grid is *pervasive*: electricity is available essentially everywhere, and one can simply access it through a standard wall-mounted socket.	The grid is also *pervasive* in the sense that the remote computing resources would be accessible from different platforms, including laptops and mobile phones, and one can simply access the grid computing power through the web browser.
The power grid is a *utility*: we ask for electricity and we get it. We also pay for what we get.	The grid computing is also a *utility*: we ask for computing power or storage capacity and we get it. We also pay for what we get.

Grid computing is more popular due to the following reasons:

- Its ability to make use of unused computing power, and thus, it is a cost-effective solution (reducing investments, only recurring costs)
- As a way to solve problems in line with any HPC-based application
- Enables heterogeneous resources of computers to work cooperatively and collaboratively to solve a scientific problem

Researchers associate the term *grid* to the way electricity is distributed in municipal areas for the common man. In this context, the difference between electrical power grid and grid computing is worth noting (Table 1.1).

1.6 Cloud Computing

The computing trend moved toward cloud from the concept of grid computing, particularly when large computing resources are required to solve a single problem, using the ideas of computing power as a *utility* and other allied concepts. However, the potential difference between grid and cloud is that grid computing supports leveraging several computers in parallel to solve a particular application, while cloud computing supports leveraging multiple resources, including computing resources, to deliver a unified *service* to the end user.

In cloud computing, the IT and business resources, such as servers, storage, network, applications, and processes, can be dynamically provisioned to the user needs and workload. In addition, while a cloud can provision and support a grid, a cloud can also support nongrid environments, such as a three-tier web architecture running on traditional or Web 2.0 applications.

We will be looking at the details of cloud computing in different chapters of this book.

1.7 Biocomputing

Biocomputing systems use the concepts of biologically derived or simulated molecules (or models) that perform computational processes in order to solve a problem. The biologically derived models aid in structuring the computer programs that become part of the application.

Biocomputing provides the theoretical background and practical tools for scientists to explore proteins and DNA. DNA and proteins are nature's building blocks, but these building blocks are not exactly used as *bricks*; the function of the final molecule rather strongly depends on the *order* of these blocks. Thus, the biocomputing scientist works on inventing the *order* suitable for various applications mimicking biology. Biocomputing shall, therefore, lead to a better understanding of life and the molecular causes of certain diseases.

1.8 Mobile Computing

In mobile computing, the processing (or computing) elements are small (i.e., handheld devices) and the communication between various resources is taking place using wireless media.

Mobile communication for voice applications (e.g., cellular phone) is widely established throughout the world and witnesses a very rapid growth in all its dimensions including the increase in the number of subscribers of various cellular networks. An extension of this technology is the ability to send and receive data across various cellular networks using small devices such as smartphones. There can be numerous applications based on this technology; for example, video call or conferencing is one of the important applications that people prefer to use in place of existing voice (only) communications on mobile phones.

Mobile computing–based applications are becoming very important and rapidly evolving with various technological advancements as it allows users to transmit data from remote locations to other remote or fixed locations.

1.9 Quantum Computing

Manufacturers of computing systems say that there is a limit for cramming more and more transistors into smaller and smaller spaces of integrated circuits (ICs) and thereby doubling the processing power about every 18 months. This problem will have to be overcome by a new *quantum computing*–based solution, wherein the dependence is on quantum information, the rules that govern the subatomic world. Quantum computers are millions of times faster than even our most powerful supercomputers today. Since quantum computing works differently on the most fundamental level than the current technology, and although there are working prototypes, these systems have not so far proved to be alternatives to today's silicon-based machines.

1.10 Optical Computing

Optical computing system uses the photons in visible light or infrared beams, rather than electric current, to perform digital computations. An electric current flows at only about 10% of the speed of light. This limits the rate at which data can be exchanged over long distances and is one of the factors that led to the evolution of optical fiber. By applying some of the advantages of visible and/or IR networks at the device and component scale, a computer can be developed that can perform operations 10 or more times faster than a conventional electronic computer.

1.11 Nanocomputing

Nanocomputing refers to computing systems that are constructed from nanoscale components. The silicon transistors in traditional computers may be replaced by transistors based on carbon nanotubes.

The successful realization of nanocomputers relates to the scale and integration of these nanotubes or components. The issues of scale relate to the dimensions of the components; they are, at most, a few nanometers in at least two dimensions. The issues of integration of the components are twofold: first, the manufacture of complex arbitrary patterns may be economically infeasible, and second, nanocomputers may include massive quantities of devices. Researchers are working on all these issues to bring nanocomputing a reality.

1.12 Network Computing

Network computing is a way of designing systems to take advantage of the latest technology and maximize its positive impact on business solutions and their ability to serve their customers using a strong underlying network of computing resources. In any network computing solution, the client component of a networked architecture or application will be with the customer or client or end user, and in modern days, they provide an essential set of functionality necessary to support the appropriate client functions at minimum cost and maximum simplicity. Unlike conventional PCs, they do not need to be individually configured and maintained according to their intended use. The other end of the client component in the network architecture will be a typical *server* environment to *push* the services of the application to the client end.

Almost all the computing paradigms that were discussed earlier are of this nature. Even in the future, if any one invents a totally new computing paradigm, it would be based on a networked architecture, without which it is impossible to realize the benefits for any end user.

1.13 Summary

We are into a post-PC era, in which a greater number and a variety of computers and computing paradigms with different sizes and functions might be used everywhere and with every human being; so, the purpose of this chapter is to illustrate briefly the ideas of all these computing domains, as most of these are ubiquitous and pervasive in its access and working environment.

Key Points

- *Mobile computing*: Mobile computing consists of small processing elements (i.e., handheld devices) and the communication between various resources is by using wireless media (see Section 1.8).
- *Nanocomputing*: Makes use of nanoscale components (see Section 1.11).

Review Questions

1. Why is it necessary to understand the various computing paradigms?
2. Compare grid computing with electric power grid
3. Will mobile computing play a dominant role in the future? Discuss
4. How are distributed computing and network computing different or similar?
5. How may nanocomputing shape future devices?

Further Reading

Ditto, W. L., A. Miliotis, K. Murali, and S. Sinha. The chaos computing paradigm. *Reviews of Nonlinear Dynamics and Complexity* 3: 1–35, 2010.

2

Cloud Computing Fundamentals

Learning Objectives

The objectives of this chapter are to

- Understand the basic ideas and motivation for cloud computing
- To define cloud computing
- Understand the 5-4-3 principles of cloud computing and cloud ecosystem
- Understand the working of a cloud application
- Have a brief understanding on the benefits and drawbacks in cloud computing

Preamble

Modern computing with our laptop or desktop or even with tablets/smart-phones using the Internet to access the data and details that we want, which are located/stored at remote places/computers, through the faces of applications like Facebook, e-mail, and YouTube, brings the actual power of information that we need instantaneously within no time. Even if millions of users get connected in this manner, from anywhere in the world, these applications do serve what these users–customers want. This phenomenon of supply of information or any other data and details to all the needy customers, as and when it is asked, is the conceptual understanding and working of what is known as cloud computing. This chapter is devoted to give basic understanding on cloud computing.

2.1 Motivation for Cloud Computing

Let us review the scenario of computing prior to the announcement and availability of cloud computing: The users who are in need of computing are expected to invest money on computing resources such as hardware, software, networking, and storage; this investment naturally costs a bulk currency to the users as they have to buy these computing resources, keep these in their premises, and maintain and make it operational—all these tasks would add cost. And, this is a particularly true and huge expenditure to the enterprises that require enormous computing power and resources, compared with classical academics and individuals.

On the other hand, it is easy and handy to get the required computing power and resources from some provider (or supplier) as and when it is needed and pay only for that usage. This would cost only a reasonable investment or spending, compared to the huge investment when buying the entire computing infrastructure. This phenomenon can be viewed as *capital expenditure* versus *operational expenditure*. As one can easily assess the huge lump sum required for capital expenditure (whole investment and maintenance for computing infrastructure) and compare it with the moderate or smaller lump sum required for the hiring or getting the computing infrastructure only to the tune of required time, and rest of the time free from that. Therefore, cloud computing is a mechanism of *bringing–hiring or getting the services of the computing power or infrastructure* to an organizational or individual level to the extent required and paying only for the consumed services.

One can compare this situation with the usage of electricity (its services) from its producer-cum-distributor (in India, it is the state-/government-owned electricity boards that give electricity supply to all residences and organizations) to houses or organizations; here, we do not generate electricity (comparable with electricity production–related tasks); rather, we use it only to tune up our requirements in our premises, such as for our lighting and usage of other electrical appliances, and pay as per the electricity meter reading value.

Therefore, cloud computing is needed in getting the services of computing resources. Thus, one can say as a one-line answer to the need for cloud computing that it eliminates a large computing investment without compromising the use of computing at the user level at an operational cost. Cloud computing is very economical and saves a lot of money. A blind benefit of this computing is that even if we lose our laptop or due to some crisis our personal computer—and the desktop system—gets damaged, still our data and files will stay safe and secured as these are not in our local machine (but remotely located at the provider's place—machine).

In addition, one can think to add security while accessing these remote computing resources as depicted in Figure 2.1.

Figure 2.1 shows several cloud computing applications. The *cloud* represents the Internet-based computing resources, and the accessibility is through some

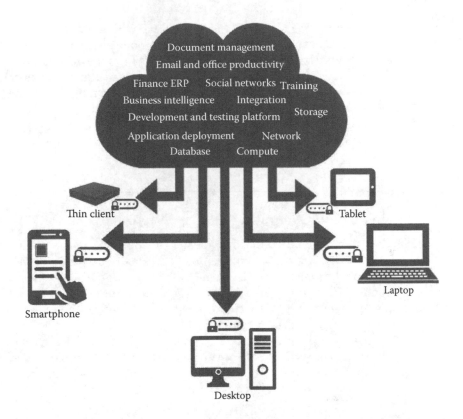

FIGURE 2.1
Cloud computing.

secure support of connectivity. It is a computing solution growing in popularity, especially among individuals and small- and medium-sized companies (SMEs). In the cloud computing model, an organization's core computer power resides offsite and is essentially subscribed to rather than owned.

Thus, cloud computing comes into focus and much needed only when we think about what computing resources and information technology (IT) solutions are required. This need caters to a way to increase capacity or add capabilities on the fly without investing in new infrastructure, training new personnel, or licensing new software. Cloud computing encompasses the subscription-based or pay-per-use service model of offering computing to end users or customers over the Internet and thereby extending the IT's existing capabilities.

2.1.1 The Need for Cloud Computing

The main reasons for the need and use of cloud computing are convenience and reliability. In the past, if we wanted to bring a file, we would have to save it to a Universal Serial Bus (USB) flash drive, external hard drive, or compact disc (CD) and bring that device to a different place. Instead, saving a file to the cloud (e.g., use of cloud application Dropbox) ensures that we will be able to access it with any computer that has an Internet connection. The cloud also makes it much easier to share a file with friends, making it possible to collaborate over the web.

While using the cloud, losing our data/file is much less likely. However, just like anything online, there is always a risk that someone may try to gain access to our personal data, and therefore, it is important to choose an access control with a strong password and pay attention to any privacy settings for the cloud service that we are using.

2.2 Defining Cloud Computing

In the simplest terms, cloud computing means storing and accessing data and programs over the Internet from a remote location or computer instead of our computer's hard drive. This so called *remote location* has several properties such as scalability, elasticity etc., which is significantly different from a simple remote machine. The cloud is just a metaphor for the Internet. When we store data on or run a program from the local computer's hard drive, that is called local storage and computing. For it to be considered *cloud computing*, we need to access our data or programs over the Internet. The end result is the same; however, with an online connection, cloud computing can be done anywhere, anytime, and by any device.

2.2.1 NIST Definition of Cloud Computing

The formal definition of cloud computing comes from the National Institute of Standards and Technology (NIST): "Cloud computing is a model for enabling ubiquitous, convenient, on-demand network access to a shared pool of configurable computing resources (e.g., networks, servers, storage, applications, and services) that can be rapidly provisioned and released with minimal management effort or service provider interaction. This cloud model is composed of five essential characteristics, three service models, and four deployment models [1].

It means that the computing resource or infrastructure—be it server hardware, storage, network, or application software—all available from the cloud vendor or provider's site/premises, can be accessible over the Internet from any remote location and by any local computing device. In addition, the usage or accessibility is to cost only to the level of usage to the customers based on their needs and demands, also known as the *pay-as-you-go* or *pay-as-per-use* model. If the need is more, more quantum computing resources are made available (provisioning with elasticity) by the provider. Minimal management effort implies that at the customer's side, the maintenance of computing systems is very minimal as they will have to look at these tasks only for their local computing devices used for accessing cloud-based resources, not for those computing resources managed at the provider's side. Details of five essential characteristics, three service models, and four deployment models are provided in the 5-4-3 principles in Section 2.3. Many vendors, pundits, and experts refer to NIST, and both the International Standards Organization (ISO) and the Institute of Electrical and Electronics Engineers (IEEE) back the NIST definition.

Now, let us try to define and understand cloud computing from two other perspectives—as a service and a platform—in the following sections.

2.2.2 Cloud Computing Is a Service

The simplest thing that any computer does is allow us to store and retrieve information. We can store our family photographs, our favorite songs, or even save movies on it, which is also the most basic service offered by cloud computing. Let us look at the example of a popular application called *Flickr* to illustrate the meaning of this section.

While Flickr started with an emphasis on sharing photos and images, it has emerged as a great place to store those images. In many ways, it is superior to storing the images on your computer:

1. First, Flickr allows us to easily access our images no matter where we are or what type of device we are using. While we might upload the photos of our vacation from our home computer, later, we can easily access them from our laptop at the office.

2. Second, Flickr lets us share the images. There is no need to burn them to a CD or save them on a flash drive. We can just send someone our Flickr address to share these photos or images.

3. Third, Flickr provides data security. By uploading the images to Flickr, we are providing ourselves with data security by creating a backup on the web. And, while it is always best to keep a local copy—either on a computer, a CD, or a flash drive—the truth is that we are far more likely to lose the images that we store locally than Flickr is of losing our images.

2.2.3 Cloud Computing Is a Platform

The World Wide Web (WWW) can be considered as the operating system for all our Internet-based applications. However, one has to understand that we will always need a local operating system in our computer to access web-based applications.

The basic meaning of the term *platform* is that it is the support on which applications run or give results to the users. For example, Microsoft Windows is a platform. But, a platform does not have to be an operating system. Java is a platform even though it is not an operating system.

Through cloud computing, the web is becoming a platform. With trends (applications) such as Office 2.0, more and more applications that were originally available on desktop computers are now being converted into web–cloud applications. Word processors like Buzzword and office suites like Google Docs are now available in the cloud as their desktop counterparts. All these kinds of trends in providing applications via the cloud are turning cloud computing into a platform or to act as a platform.

2.3 5-4-3 Principles of Cloud computing

The 5-4-3 principles put forth by NIST describe (a) the five essential characteristic features that promote cloud computing, (b) the four deployment models that are used to narrate the cloud computing opportunities for customers while looking at architectural models, and (c) the three important and basic service offering models of cloud computing.

2.3.1 Five Essential Characteristics

Cloud computing has five essential characteristics, which are shown in Figure 2.2. Readers can note the word *essential*, which means that if any of these characteristics is missing, then it is not cloud computing:

1. *On-demand self-service*: A consumer can unilaterally provision computing capabilities, such as server time and network storage, as needed automatically without requiring human interaction with each service's provider.
2. *Broad network access*: Capabilities are available over the network and accessed through standard mechanisms that promote use by heterogeneous thin or thick client platforms (e.g., mobile phones, laptops, and personal digital assistants [PDAs]).

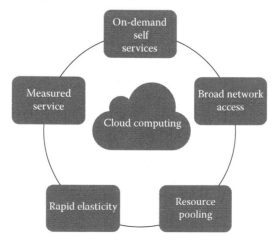

FIGURE 2.2
The essential characteristics of cloud computing.

3. *Elastic resource pooling*: The provider's computing resources are pooled to serve multiple consumers using a multitenant model, with different physical and virtual resources dynamically assigned and reassigned according to consumer demand. There is a sense of location independence in that the customer generally has no control or knowledge over the exact location of the provided resources but may be able to specify the location at a higher level of abstraction (e.g., country, state, or data center). Examples of resources include storage, processing, memory, and network bandwidth.

4. *Rapid elasticity*: Capabilities can be rapidly and elastically provisioned, in some cases automatically, to quickly scale out and rapidly released to quickly scale in. To the consumer, the capabilities available for provisioning often appear to be unlimited and can be purchased in any quantity at any time.

5. *Measured service*: Cloud systems automatically control and optimize resource use by leveraging a metering capability at some level of abstraction appropriate to the type of service (e.g., storage, processing, bandwidth, and active user accounts). Resource usage can be monitored, controlled, and reported providing transparency for both the provider and consumer of the utilized service.

2.3.2 Four Cloud Deployment Models

Deployment models describe the ways with which the cloud services can be deployed or made available to its customers, depending on the organizational structure and the provisioning location. One can understand it in this manner too: cloud (Internet)-based computing resources—that is, the locations where data and services are acquired and provisioned to its customers—can take various forms. Four deployment models are usually distinguished, namely, public, private, community, and hybrid cloud service usage:

1. *Private cloud*: The cloud infrastructure is provisioned for exclusive use by a single organization comprising multiple consumers (e.g., business units). It may be owned, managed, and operated by the organization, a third party, or some combination of them, and it may exist on or off premises.

2. *Public cloud*: The cloud infrastructure is provisioned for open use by the general public. It may be owned, managed, and operated by a business, academic, or government organization, or some combination of them. It exists on the premises of the cloud provider.

3. *Community cloud*: The cloud infrastructure is shared by several organizations and supports a specific community that has shared concerns (e.g., mission, security requirements, policy, and compliance considerations). It may be managed by the organizations or a third party and may exist on premise or off premise.

4. *Hybrid cloud*: The cloud infrastructure is a composition of two or more distinct cloud infrastructures (private, community, or public) that remain unique entities but are bound together by standardized or proprietary technology that enables data and application portability (e.g., cloud bursting for load balancing between clouds).

2.3.3 Three Service Offering Models

The three kinds of services with which the cloud-based computing resources are available to end customers are as follows: Software as a Service (SaaS), Platform as a Service (PaaS), and Infrastructure as a Service (IaaS). It is also known as the service–platform–infrastructure (SPI) model of the cloud and is shown in Figure 2.3. SaaS is a software distribution model in which applications (software, which is one of the most important computing resources) are hosted by a vendor or service provider and made available to customers over a network, typically the Internet. PaaS is a paradigm for delivering operating systems and associated services (e.g., computer aided software engineering [CASE] tools, integrated development environments [IDEs] for developing software solutions) over the Internet without downloads or installation. IaaS involves outsourcing the equipment used to support operations, including storage, hardware, servers, and networking components.

1. *Cloud SaaS*: The capability provided to the consumer is to use the provider's applications running on a cloud infrastructure, including network, servers, operating systems, storage, and even individual application capabilities, with the possible exception of limited user-specific application configuration settings. The applications are accessible from various client devices through either a thin client

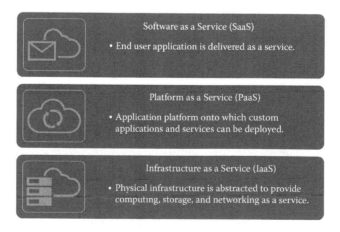

FIGURE 2.3
SPI—service offering model of the cloud.

interface, such as a web browser (e.g., web-based e-mail), or a program interface. The consumer does not manage or control the underlying cloud infrastructure. Typical applications offered as a service include customer relationship management (CRM), business intelligence analytics, and online accounting software.

2. *Cloud PaaS*: The capability provided to the consumer is to deploy onto the cloud infrastructure consumer-created or acquired applications created using programming languages, libraries, services, and tools supported by the provider. The consumer does not manage or control the underlying cloud infrastructure but has control over the deployed applications and possibly configuration settings for the application-hosting environment. In other words, it is a packaged and ready-to-run development or operating framework. The PaaS vendor provides the networks, servers, and storage and manages the levels of scalability and maintenance. The client typically pays for services used. Examples of PaaS providers include Google App Engine and Microsoft Azure Services.

3. *Cloud IaaS*: The capability provided to the consumer is to provision processing, storage, networks, and other fundamental computing resources on a pay-per-use basis where he or she is able to deploy and run arbitrary software, which can include operating systems and applications. The consumer does not manage or control the underlying cloud infrastructure but has control over the operating systems, storage, and deployed applications and possibly limited control of select networking components (e.g., host firewalls). The service provider owns the equipment and is responsible for housing, cooling operation, and maintenance. Amazon Web Services (AWS) is a popular example of a large IaaS provider.

The major difference between PaaS and IaaS is the amount of control that users have. In essence, PaaS allows vendors to manage everything, while IaaS requires more management from the customer side. Generally speaking, organizations that already have a software package or application for a specific purpose and want to install and run it in the cloud should opt to use IaaS instead of PaaS.

2.4 Cloud Ecosystem

Cloud ecosystem is a term used to describe the complete environment or system of interdependent components or entities that work together to enable and support the cloud services. To be more precise, the cloud computing's ecosystem is a complex environment that includes the description of every item or entity along with their interaction; the complex entities include the traditional elements of cloud computing such as software (SaaS), hardware (PaaS and/or IaaS), other infrastructure (e.g., network, storage), and also stakeholders like consultants, integrators, partners, third parties, and anything in their environments that has a bearing on the other components of the cloud.

The cloud ecosystem of interacting components and organizations with individuals, together known as the actors who could be responsible for either providing or consuming cloud services, can be categorized in the following manner:

1. *Cloud service users (CSUs)*: A consumer (an individual/person), enterprise (including enterprise administrator), and/or government/public institution or organization that consumes delivered cloud services; a CSU can include intermediate users that will deliver cloud services provided by a cloud service provider (CSP) to actual users of the cloud service, that is, end users. End users can be persons, machines, or applications.

2. *CSPs*: An organization that provides or delivers and maintains or manages cloud services, that is, provider of SaaS, PaaS, IaaS, or any allied computing infrastructure.

3. *Cloud service partners (CSNs)*: A person or organization (e.g., application developer; content, software, hardware, and/or equipment provider; system integrator; and/or auditor) that provides support to the building of a service offered by a CSP (e.g., service integration).

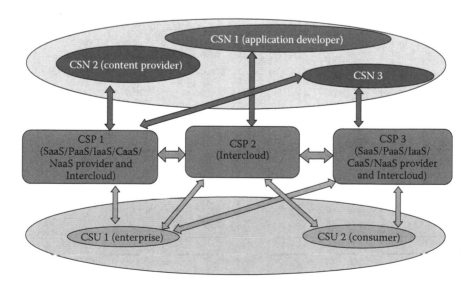

FIGURE 2.4
Actors with some of their possible roles in a cloud ecosystem.

In layman's terms, the cloud ecosystem describes the usage and value of each entity in the ecosystem, and when all the entities in the ecosystem are put together, users are now able to have an integrated suite made up of the best-of-breed solutions. An example of this ecosystem can be a cloud accounting solution such as *Tally*; while this SaaS vendor focuses on their support for accounting and integrated payroll solutions, they can engage (collaborate) with any other third-party CSPs who could support additional features in the accounting software like reporting tools, dashboards, work papers, workflow, project management, and CRM, covering the majority of a client or customer firm's software needs. And, any other additional requirement that may be essential will likely be added by a partner joining the ecosystem in the near future. Figure 2.4 illustrates the idea of a cloud ecosystem.

2.5 Requirements for Cloud Services

From the concepts illustrated in the earlier sections, one can understand that the cloud services or service offering models require certain features to be exhibited in order to be considered as *services*. The following are the basic requirements for anything that can be considered as a service by the actors of the cloud computing ecosystem, which can be offered or provisioned through the cloud:

1. *Multitenancy*: Multitenancy is an essential characteristic of cloud systems aiming to provide isolation of the different users of the cloud system (tenants) while maximizing resource sharing. It is expected that multitenancy be supported at various levels of a cloud infrastructure. As an example, at the application level, multitenancy is a feature that allows a single instance of an application (say, database system) and leverages the economy of scale to satisfy several users at the same time.

2. *Service life cycle management*: Cloud services are paid as per usage and can be started and ended at any time. Therefore, it is required that a cloud service support automatic service provisioning. In addition, metering and charging or billing settlement needs to be provided for services that are dynamically created, modified, and then released in virtual environments.

3. *Security*: The security of each individual service needs to be protected in the multitenant cloud environment; the users (tenants) also support the needed secured services, meaning that a cloud provides strict control for tenants' service access to different resources to avoid the abuse of cloud resources and to facilitate the management of CSUs by CSPs.

4. *Responsiveness*: The cloud ecosystem is expected to enable early detection, diagnosis, and fixing of service-related problems in order to help the customers use the services faithfully.

5. *Intelligent service deployment*: It is expected that the cloud enables efficient use of resources in service deployment, that is, maximizing the number of deployed services while minimizing the usage of resources and still respecting the SLAs. For example, the specific application characteristics (e.g., central processing unit [CPU]-intensive, input/output [IO]-intensive) that can be provided by developers or via application monitoring may help CSPs in making efficient use of resources.

6. *Portability*: It is expected that a cloud service supports the portability of its features over various underlying resources and that CSPs should be able to accommodate cloud workload portability (e.g., VM portability) with limited service disruption.

7. *Interoperability*: It is expected to have available well-documented and well-tested specifications that allow heterogeneous systems in cloud environments to work together.

8. *Regulatory aspects*: All applicable regulations shall be respected, including privacy protection.

9. *Environmental sustainability*: A key characteristic of cloud computing is the capability to access, through a broad network and thin clients, on-demand shared pools of configurable resources that can be rapidly provisioned and released. Cloud computing can then be considered in its essence as an ICT energy consumption consolidation model, supporting mainstream technologies aiming to optimize energy consumption (e.g., in data centers) and application performance. Examples of such technologies include virtualization and multitenancy.

10. *Service reliability, service availability, and quality assurance*: CSUs demand for their services end-to-end quality of service (QoS) assurance, high levels of reliability, and continued availability to their CSPs.

11. *Service access*: A cloud infrastructure is expected to provide CSUs with access to cloud services from any user device. It is expected that CSUs have a consistent experience when accessing cloud services.

12. *Flexibility*: It is expected that the cloud service be capable of supporting multiple cloud deployment models and cloud service categories.

13. *Accounting and charging*: It is expected that a cloud service be capable to support various accounting and charging models and policies.

14. *Massive data processing*: It is expected that a cloud supports mechanisms for massive data processing (e.g., extracting, transforming, and loading data). It is worth to note in this context that distributed and/

or parallel processing systems will be used in cloud infrastructure deployments to provide large-scale integrated data storage and processing capabilities that scale with software-based fault tolerance.

The expected requirements for services in the IaaS category include the following:

- Computing hardware requirements (including processing, memory, disk, network interfaces, and virtual machines)
- Computing software requirements (including OS and other preinstalled software)
- Storage requirements (including storage capacity)
- Network requirements (including QoS specifications, such as bandwidth and traffic volumes)
- Availability requirements (including protection/backup plan for computing, storage, and network resources)

The expected service requirements for services in the PaaS category include the following:

- Requirements similar to those of the IaaS category
- Deployment options of user-created applications (e.g., scale-out options)

The expected service requirements for services in the SaaS category include the following:

- Application-specific requirements (including licensing options)
- Network requirements (including QoS specifications such as bandwidth and traffic volumes)

2.6 Cloud Application

A cloud application is an application program that functions or executes in the cloud; the application can exhibit some characteristics of a pure desktop application and some characteristics of a pure web-based application. A desktop application resides entirely on a single device at the user's location (it does not necessarily have to be a desktop computer), and on the other hand, a web application is stored entirely on a remote server and is delivered over the Internet through a browser interface.

Like desktop applications, cloud applications can provide fast responsiveness and can work offline. Like web applications, cloud applications need not permanently reside on the local device, but they can be easily updated online. Cloud applications are, therefore, under the user's constant control, yet they need not always consume storage space on the user's computer or communications device. Assuming that the user has a reasonably fast Internet connection, a well-written cloud application offers all the interactivity of a desktop application along with the portability of a web application.

A cloud application can be used with a web browser connected to the Internet. Now, it is possible for the user interface portion of the application to exist on the local device and for the user to cache data locally, enabling full offline mode when desired. Also, a cloud application, unlike a web app, can be used in any sensitive situation where wireless devices—connectivity—are not allowed (i.e., even when no Internet connection is available for some period).

An example of cloud application is a web-based e-mail (e.g., Gmail, Yahoo mail); in this application, the user of the e-mail uses the cloud—all of the emails in their inbox are stored on servers at remote locations at the e-mail service provider.

However, there are many other services that use the cloud in different ways. Here is yet another example: Dropbox is a cloud storage service that lets us easily store and share files with other people and access files from a mobile device as well.

2.7 Benefits and Drawbacks

One of the attractions of cloud computing is accessibility. If our applications and documents are in the cloud and are not saved on an office server, then we can access and use them at anytime, anywhere for our working, whether we are at work, at home, or even at a friend's house. Cloud computing also enables precisely the right amount of computing power and resources to be used for applications. Cloud computing vendors provide computing-related services as a bundle of computing power and parcel it out on demand. Customers can draw and make use as much or as little computing power as they need, being charged only for the usage time/computing power; accordingly, this scheme can save money. This also implies that scalability is one of the cloud computing's big benefits. When we need more computing power, cloud computing can give instant access to exactly what we need. In the cloud model, an organization's core computer power resides offsite and is essentially subscribed to rather than owned. There is no capital expenditure, only operational expenditure. It also relieves us from the responsibility and costs of maintenance of the entire computing infrastructure and pushes all these to the cloud vendor or provider. The cloud also offers a new level of reliability. The *virtualization* technology enables a vendor's cloud software to automatically move data from a piece of hardware that goes bad or is pulled offline to a section of the system or hardware that is functioning or operational. Therefore, the client gets seamless access to the data. Separate backup systems, with cloud disaster recovery strategies, provide another layer of dependability and reliability. Finally, cloud computing also promotes a *green* alternative to paper-intensive office functions. It is because it needs less computing hardware on premise, and all computing-related tasks take place remotely with minimal computing hardware requirement with the help of technological innovations such as virtualization and multitenancy. Another viewpoint on the *green* aspect is that cloud computing can reduce the environmental impact of building, shipping, housing, and ultimately destroying (or recycling) computer equipment as no one is going to own many such systems in their premises and managing the offices with fewer computers that consume less energy comparatively. A consolidated set of points briefing the benefits of cloud computing can be as follows:

1. *Achieve economies of scale*: We can increase the volume output or productivity with fewer systems and thereby reduce the cost per unit of a project or product.

2. *Reduce spending on technology infrastructure*: It is easy to access data and information with minimal upfront spending in a *pay-as-you-go* approach, in the sense that the usage and payment are similar to an electricity meter reading in the house, which is based on demand.

3. *Globalize the workforce*: People worldwide can access the cloud with Internet connection.

4. *Streamline business processes*: It is possible to get more work done in less time with less resource.

5. *Reduce capital costs*: There is no need to spend huge money on hardware, software, or licensing fees.

6. *Pervasive accessibility*: Data and applications can be accessed anytime, anywhere, using any smart computing device, making our life so much easier.

7. *Monitor projects more effectively*: It is possible to confine within budgetary allocations and can be ahead of completion cycle times.

8. *Less personnel training is needed*: It takes fewer people to do more work on a cloud, with a minimal learning curve on hardware and software issues.

9. *Minimize maintenance and licensing software*: As there is no too much of on-premise computing resources, maintenance becomes simple and updates and renewals of software systems rely on the cloud vendor or provider.

10. *Improved flexibility*: It is possible to make fast changes in our work environment without serious issues at stake.

Drawbacks to cloud computing are obvious. The main point in this context is that if we lose our Internet connection, we have lost the link to the cloud and thereby to the data and applications. There is also a concern about security as our entire working with data and applications depend on other's (cloud vendor or providers) computing power. Also, while cloud computing supports scalability (i.e., quickly scaling up and down computing resources depending on the need), it does not permit the control on these resources as these are not owned by the user or customer. Depending on the cloud vendor or provider, customers may face restrictions on the availability of applications, operating systems, and infrastructure options. And, sometimes, all development platforms may not be available in the cloud due to the fact that the cloud vendor may not aware of such solutions. A major barrier to cloud computing is the interoperabebility of applications, which is the ability of two or more applications that are required to support a business need to work together by sharing data and other business-related resources. Normally, this does not happen in the cloud as these applications may not be available with a single cloud vendor and two different vendors having these applications do not cooperate with each other.

2.8 Summary

For a clear understanding of cloud computing, there are certain fundamental concepts to be known, as discussed in this chapter. This chapter starts with the motivation for cloud computing and discusses in brief the reason for which cloud was introduced, the need for cloud computing, and the basic definition of cloud. NIST provides a standard definition for cloud computing. Cloud is based on the 5-4-3 principle. Cloud has different environments. And so, the cloud ecosystem is discussed, which briefly points out different roles involved in cloud computing. Further several essential features of cloud computing are elaborated. Applications in cloud are also briefly discussed. The chapter ends with a detailed note on the benefits and drawbacks of cloud.

Review Points

- *Cloud computing*: Cloud computing is a model for enabling ubiquitous, convenient, on-demand network access to a shared pool of configurable computing resources (e.g., networks, servers, storage, applications, and services) that can be rapidly provisioned and released with minimal management effort or service provider interaction. This cloud model is composed of five essential characteristics, three service models, and four deployment models (see Section 2.2.1).

- *Cloud ecosystem*: A person or organization (e.g., application developer; content, software, hardware, and/or equipment provider; system integrator; and/or auditor) that provides support to the building of a service offered by a CSP (e.g., service integration) (see Section 2.4).
- *Cloud service providers*: An organization that provides or delivers and maintains or manages cloud services, that is, provider of SaaS, PaaS, IaaS, or any allied computing infrastructure (see Section 2.4).
- *Multitenancy*: Multitenancy is an essential characteristic of cloud systems aiming to provide isolation of the different users of the cloud system (tenants) while maximizing resource sharing (see Section 2.5).

Review Questions

1. What is cloud computing? Why is it needed?
2. Describe a real-life example to illustrate the concepts behind cloud computing.
3. Distinguish between the definitions of *cloud computing is a service* and *cloud computing is a platform*.
4. Is it true that all essential characteristic features of the cloud are necessary to completely describe it?
5. What are the service offering models of the cloud?
6. What are the deployment models of the cloud?
7. What are the actors and their roles in a typical cloud ecosystem?
8. Enlist and explain the requirements that need to be considered for cloud services.
9. Explain how a cloud application is being accessed.
10. Give a brief note on the merits and demerits of cloud computing.

Reference

1. Mell, P. and T. Grance. The NIST definition of cloud computing. NIST Special Publication 800-145, 2011. Available [Online]: http://csrc.nist.gov/publications/nistpubs/800-145/SP800-145.pdf. Accessed September 3, 2013.

Further Reading

A complete history of cloud computing. Available [Online]: http://www.salesforce.com/
    uk/socialsuccess/cloud-computing/the-complete-history-of-cloud-computing.
    jsp. Accessed February 4, 2014.
Cloud computing for business: What is cloud. Available [Online]: http://www.
    opengroup.org/cloud/cloud/cloud_for_business/what.htm. Accessed March
    2, 2014.
Mell, P. and T. Grance. The NIST definition of cloud computing. NIST Special
    Publication 800-145, 2011. Available [Online]: http://csrc.nist.gov/publications/
    nistpubs/800-145/SP800-145.pdf. Accessed September 3, 2013.
Nations, D. What is Flickr?. Available [Online]: http://webtrends.about.com/od/
    profile1/fr/what-is-Flickr.htm. Accessed October 8, 2013.
Strikland, J. Cloud computing architecture. Available [Online]: http://computer.how-
    stuffworks.com/cloud-computing/cloud-computing1.htm. Accessed January
    8, 2014.
Ward, S. Why cloud computing is ideal for small businesses. Available [Online]: http://
    sbinfocanada.about.com/od/itmanagement/a/Why-Cloud-Computing.htm.
    Accessed March 15, 2014.
What cloud computing really means. Available [Online]: http://www.infoworld.com/d/
    cloud-computing/what-cloud-computing-really-means-031?page=0,1.
What is cloud computing?—The complete guide. Available [Online]: http://www.
    salesforce.com/uk/socialsuccess/cloud-computing/what-is-cloud-computing.jsp.
    Accessed October 28, 2014.

3

Cloud Computing Architecture and Management

Learning Objectives

The objectives of this chapter are to

- Provide an overview of the cloud architecture
- Give an insight on the anatomy of the cloud
- Describe the role of network connectivity in the cloud
- Give a description about applications in the cloud
- Give a detailed description about managing the cloud
- Provide an overview about application migration to the cloud

Preamble

Cloud computing is an emerging technology that has become one of the most popular computing technologies. Each and every technology has certain concepts that form the basis for its working. Similarly, there are several aspects of a technology that needs to be looked upon before delving deeper. Thus, there are some basic issues in cloud computing that need to be discussed before going into a detailed discussion about the cloud. This chapter firstly describes the cloud architecture. Cloud architecture consists of a hierarchical set of components that collectively describe the way the cloud works. The next section explains about the cloud anatomy, followed by network connectivity in the cloud and then the fine details about managing a cloud application. Finally, an overview on migrating applications to the cloud is discussed. Some of the topics that are discussed in this chapter are elaborated in upcoming chapters.

3.1 Introduction

Cloud computing is similar to other technologies in a way that it also has several basic concepts that one should learn before knowing its core concepts. There are several processes and components of cloud computing that need to be discussed. One of the topics of such prime importance is architecture. Architecture is the hierarchical view of describing a technology. This usually includes the components over which the existing technology is built and the components that are dependent on the technology. Another topic that is related to architecture is anatomy. Anatomy describes the core structure of the cloud. Once the structure of the cloud is clear, the network connections in the cloud and the details about the cloud application need to be known. This is important as the cloud is a completely Internet-dependent technology. Similarly, cloud management discusses the important management issues and ways in which the current cloud scenario is managed. It describes the way an application and infrastructure in the cloud are managed. Management is important because of the quality of service (QoS) factors that are involved in the cloud. These QoS factors form the basis for cloud computing. All the services are given based on these QoS factors. Similarly, application migration to the cloud also plays a very important role. Not all applications can be directly deployed to the cloud. An application needs to be properly migrated to the cloud to be considered a proper cloud application that will have all the properties of the cloud.

3.2 Cloud Architecture

Any technological model consists of an architecture based on which the model functions, which is a hierarchical view of describing the technology. The cloud also has an architecture that describes its working mechanism. It includes the dependencies on which it works and the components that work over it. The cloud is a recent technology that is completely dependent on the Internet for its functioning. Figure 3.1 depicts the architecture. The cloud architecture can be divided into four layers based on the access of the cloud by the user. They are as follows.

3.2.1 Layer 1 (User/Client Layer)

This layer is the lowest layer in the cloud architecture. All the users or client belong to this layer. This is the place where the client/user initiates the

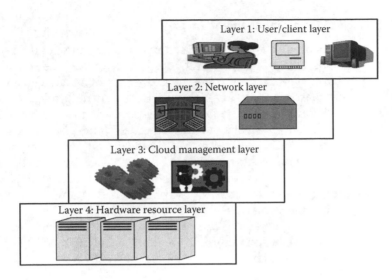

FIGURE 3.1
Cloud architecture.

connection to the cloud. The client can be any device such as a thin client, thick client, or mobile or any handheld device that would support basic functionalities to access a web application. The thin client here refers to a device that is completely dependent on some other system for its complete functionality. In simple terms, they have very low processing capability. Similarly, thick clients are general computers that have adequate processing capability. They have sufficient capability for independent work. Usually, a cloud application can be accessed in the same way as a web application. But internally, the properties of cloud applications are significantly different. Thus, this layer consists of client devices.

3.2.2 Layer 2 (Network Layer)

This layer allows the users to connect to the cloud. The whole cloud infrastructure is dependent on this connection where the services are offered to the customers. This is primarily the Internet in the case of a public cloud. The public cloud usually exists in a specific location and the user would not know the location as it is abstract. And, the public cloud can be accessed all over the world. In the case of a private cloud, the connectivity may be provided by a local area network (LAN). Even in this case, the cloud completely depends on the network that is used. Usually, when accessing the public or private cloud, the users require minimum bandwidth, which is sometimes defined by the cloud providers. This layer does not come under the purview of service-level agreements (SLAs), that is, SLAs do not take into account the Internet connection between the user and cloud for quality of service (QoS).

3.2.3 Layer 3 (Cloud Management Layer)

This layer consists of softwares that are used in managing the cloud. The softwares can be a cloud operating system (OS), a software that acts as an interface between the data center (actual resources) and the user, or a management software that allows managing resources. These softwares usually allow resource management (scheduling, provisioning, etc.), optimization (server consolidation, storage workload consolidation), and internal cloud governance. This layer comes under the purview of SLAs, that is, the operations taking place in this layer would affect the SLAs that are being decided upon between the users and the service providers. Any delay in processing or any discrepancy in service provisioning may lead to an SLA violation. As per rules, any SLA violation would result in a penalty to be given by the service provider. These SLAs are for both private and public clouds Popular service providers are Amazon Web Services (AWS) and Microsoft Azure for public cloud. Similarly, OpenStack and Eucalyptus allow private cloud creation, deployment, and management.

3.2.4 Layer 4 (Hardware Resource Layer)

Layer 4 consists of provisions for actual hardware resources. Usually, in the case of a public cloud, a data center is used in the back end. Similarly, in a private cloud, it can be a data center, which is a huge collection of hardware resources interconnected to each other that is present in a specific location or a high configuration system. This layer comes under the purview of SLAs. This is the most important layer that governs the SLAs. This layer affects the SLAs most in the case of data centers. Whenever a user accesses the cloud, it should be available to the users as quickly as possible and should be within the time that is defined by the SLAs. As mentioned, if there is any discrepancy in provisioning the resources or application, the service provider has to pay the penalty. Hence, the data center consists of a high-speed network connection and a highly efficient algorithm to transfer the data from the data center to the manager. There can be a number of data centers for a cloud, and similarly, a number of clouds can share a data center.

Thus, this is the architecture of a cloud. The layering is strict, and for any cloud application, this is followed. There can be a little loose isolation between layer 3 and layer 4 depending on the way the cloud is deployed.

3.3 Anatomy of the Cloud

Cloud anatomy can be simply defined as the structure of the cloud. Cloud anatomy cannot be considered the same as cloud architecture. It may not include any dependency on which or over which the technology works,

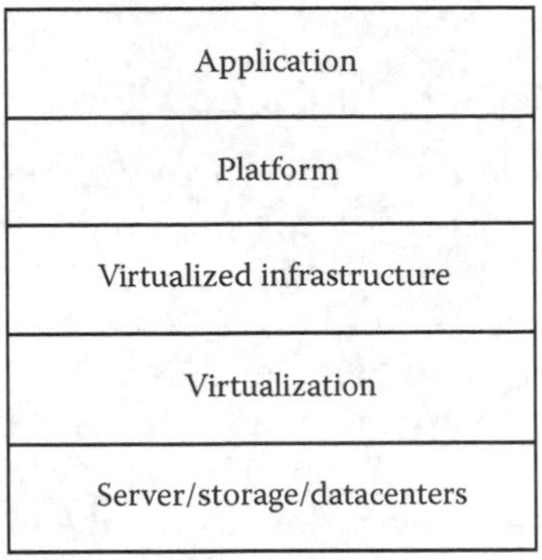

Application
Platform
Virtualized infrastructure
Virtualization
Server/storage/datacenters

FIGURE 3.2
Cloud structure.

whereas architecture wholly defines and describes the technology over which it is working. Architecture is a hierarchical structural view that defines the technology as well as the technology over which it is dependent or/and the technology that are dependent on it. Thus, anatomy can be considered as a part of architecture. The basic structure of the cloud is described in Figure 3.2, which can be elaborated, and minute structural details can be given. Figure 3.2 depicts the most standard anatomy that is the base for the cloud. It depends on the person to choose the depth of description of the cloud. A different view of anatomy is given by Refs. [1,2].

There are basically five components of the cloud:

1. *Application*: The upper layer is the application layer. In this layer, any applications are executed.
2. *Platform*: This component consists of platforms that are responsible for the execution of the application. This platform is between the infrastructure and the application.
3. *Infrastructure*: The infrastructure consists of resources over which the other components work. This provides computational capability to the user.
4. *Virtualization*: Virtualization is the process of making logical components of resources over the existing physical resources. The logical components are isolated and independent, which form the infrastructure.
5. *Physical hardware*: The physical hardware is provided by server and storage units.

These components are the basis and are described in detail in further chapters.

3.4 Network Connectivity in Cloud Computing

Cloud computing is a technique of resource sharing where servers, storage, and other computing infrastructure in multiple locations are connected by networks. In the cloud, when an application is submitted for its execution, needy and suitable resources are allocated from this collection of resources; as these resources are connected via the Internet, the users get their required results. For many cloud computing applications, network performance will be the key issue to cloud computing performance. Since cloud computing has various deployment options, we now consider the important aspects related to the cloud deployment models and their accessibility from the viewpoint of network connectivity.

3.4.1 Public Cloud Access Networking

In this option, the connectivity is often through the Internet, though some cloud providers may be able to support virtual private networks (VPNs) for customers. Accessing public cloud services will always create issues related to security, which in turn is related to performance. One of the possible approaches toward the support of security is to promote connectivity through encrypted tunnels, so that the information may be sent via secure pipes on the Internet. This procedure will be an overhead in the connectivity, and using it will certainly increase delay and may impact performance.

If we want to reduce the delay without compromising security, then we have to select a suitable routing method such as the one reducing the delay by minimizing transit *hops* in the end-to-end connectivity between the cloud provider and cloud consumer. Since the end-to-end connectivity support is via the Internet, which is a complex federation of interconnected providers (known as Internet service providers [ISPs]), one has to look at the options of selecting the path.

3.4.2 Private Cloud Access Networking

In the private cloud deployment model, since the cloud is part of an organizational network, the technology and approaches are local to the in-house network structure. This may include an Internet VPN or VPN service from a network operator. If the application access was properly done with an organizational network—connectivity in a *precloud* configuration—transition to private cloud computing will not affect the access performance.

3.4.3 Intracloud Networking for Public Cloud Services

Another network connectivity consideration in cloud computing is intracloud networking for public cloud services. Here, the resources of the cloud provider and thus the cloud service to the customer are based on the resources that are geographically apart from each other but still connected via the Internet. Public cloud computing networks are internal to the service provider and thus not visible to the user/customer; however, the security aspects of connectivity and the access mechanisms of the resources are important. Another issue to look for is the QoS in the connected resources worldwide. Most of the performance issues and violations from these are addressed in the SLAs commercially.

3.4.4 Private Intracloud Networking

The most complicated issue for networking and connectivity in cloud computing is private intracloud networking. What makes this particular issue so complex is that it depends on how much intracloud connectivity is associated with the applications being executed in this environment. Private intracloud networking is usually supported over connectivity between the major data center sites owned by the company. At a minimum, all cloud computing implementations will rely on intracloud networking to link users with the resource to which their application was assigned. Once the resource linkage is made, the extent to which intracloud networking is used depends on whether the application is componentized based on *service-oriented architecture (SOA)* or not, among multiple systems. If the principle of SOA is followed, then traffic may move between components of the application, as well as between the application and the user. The performance of those connections will then impact cloud computing performance overall. Here too, the impact of cloud computing performance is the differences that exist between the current application and the network relationships with the application.

There are reasons to consider the networks and connectivity in cloud computing with newer approaches as globalization and changing network requirements, especially those related to increased Internet usage, are demanding more flexibility in the network architectures of today's enterprises. How are these related to us? The answers are discussed later.

3.4.5 New Facets in Private Networks

Conventional private networks have been architected for on-premise applications and maximum Internet security. Typically, applications such as e-mail, file sharing, and *enterprise resource planning* (ERP) systems are delivered to on-premise-based servers at each corporate data center. Increasingly today, software vendors are offering Software as a Service (SaaS) as an alternative for their software support to the corporate offices, which brings more challenges in the access and usage mechanisms of software from data center servers and in the connectivity of network architectures. The traditional network architecture for these global enterprises was not designed to optimize performance for cloud applications, now that many applications including mission-critical applications are transitioning (moving) from on-premise based to cloud based, wherein the network availability becomes as mission critical as electricity: the business cannot function if it cannot access applications such as ERP and e-mail.

3.4.6 Path for Internet Traffic

The traditional Internet traffic through a limited set of Internet gateways poses performance and availability issues for end users who are using cloud-based applications. It can be improved if a more widely distributed Internet gateway infrastructure and connectivity are being supported for accessing applications, as they will provide lower-latency access to their cloud applications. As the volume of traffic to cloud applications grows, the percentage of the legacy network's capacity in terms of traffic to regional gateways increases. Applications such as video conferencing would hog more bandwidth while mission-critical applications such as ERP will consume less bandwidth, and hence, one has to plan a correct connectivity and path between providers and consumers.

3.5 Applications on the Cloud

The power of a computer is realized through the applications. There are several types of applications. The first type of applications that was developed and used was a stand-alone application. A stand-alone application is developed to be run on a single system that does not use network for its functioning. These stand-alone systems use only the machine in which they are installed. The functioning of these kinds of systems is totally dependent on the resources or features available within the system. These systems do not need the data or processing power of other systems; they are self-sustaining. But as the time passed, the requirements of the users changed and certain applications were required, which could be accessed by other users away from the systems. This led to the inception of web application.

The web applications were different from the stand-alone applications in many aspects. The main difference was the client server architecture that was followed by the web application. Unlike stand-alone applications, these systems were totally dependent on the network for its working. Here, there are basically two components, called as the client and the server. The server is a high-end machine that consists of the web application installed. This web application is accessed from other client systems. The client can reside anywhere in the network. It can access the web application through the Internet. This type of application was very useful, and this is extensively used from its inception and now has become an

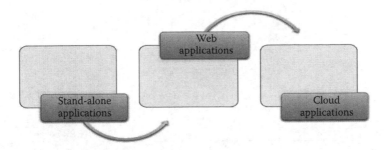

FIGURE 3.3
Computer application evolution.

important part of day-to-day life. Though this application is much used, there are shortcomings as discussed in the following:

- The web application is not elastic and cannot handle very heavy loads, that is, it cannot serve highly varying loads.
- The web application is not multitenant.
- The web application does not provide a quantitative measurement of the services that are given to the users, though they can monitor the user.
- The web applications are usually in one particular platform.
- The web applications are not provided on a pay-as-you-go basis; thus, a particular service is given to the user for permanent or trial use and usually the timings of user access cannot be monitored.
- Due to its nonelastic nature, peak load transactions cannot be handled.

Primarily to solve the previously mentioned problem, the cloud applications were developed. Figure 3.3 depicts the improvements in the applications.

The cloud as mentioned can be classified into three broad access or service models, Software as a Service (SaaS), Platform as a Service (PaaS), and Infrastructure as a Service (IaaS). Cloud application in general refers to a SaaS application.

A cloud application is different from other applications; they have unique features. A cloud application usually can be accessed as a web application but its properties differ. According to NIST [3], the features that make cloud applications unique are described in the following (Figure 3.4 depicts the features of a cloud application):

1. *Multitenancy*: Multitenancy is one of the important properties of cloud that make it different from other types of application in which the software can be shared by different users with full independence. Here, independence refers to logical independence.

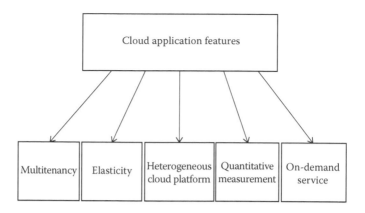

FIGURE 3.4
Features of cloud.

Each user will have a separate application instance and the changes in one application would not affect the other. Physically, the software is shared and is not independent. The degree of physical isolation is very less. The logical independence is what is guaranteed. There are no restrictions in the number of applications being shared. The difficulty in providing logical isolation depends on the physical isolation to a certain extent. If an application is physically too close, then it becomes difficult to provide multitenancy. Web application and cloud application are similar as the users use the same way to access both. Figure 3.5 depicts a multitenant application where several users share the same application.

2. *Elasticity*: Elasticity is also a unique property that enables the cloud to serve better. According to Herbst et al. [4], elasticity can be defined as the degree to which a system is able to adapt to workload changes by provisioning and deprovisioning resources in an autonomic manner such that at each point in time, the available resources match the current demand as closely as possible. Elasticity allows the cloud providers to efficiently handle the number of users, from one to

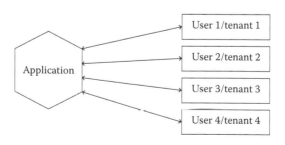

FIGURE 3.5
Multitenancy.

several hundreds of users at a time. In addition to this, it supports the rapid fluctuation of loads, that is, the increase or decrease in the number of users and their usage can rapidly change.

3. *Heterogeneous cloud platform*: The cloud platform supports heterogeneity, wherein any type of application can be deployed in the cloud. Because of this property, the cloud is flexible for the developers, which facilitates deployment. The applications that are usually deployed can be accessed by the users using a web browser.

4. *Quantitative measurement*: The services provided can be quantitatively measured. The user is usually offered services based on certain charges. Here, the application or resources are given as a utility on a pay-per-use basis. Thus, the use can be monitored and measured. Not only the services are measureable, but also the link usage and several other parameters that support cloud applications can be measured. This property of measuring the usage is usually not available in a web application and is a unique feature for cloud-based applications.

5. *On-demand service*: The cloud applications offer service to the user, on demand, that is, whenever the user requires it. The cloud service would allow the users to access web applications usually without any restrictions on time, duration, and type of device used.

The previously mentioned properties are some of the features that make cloud a unique application platform. These properties mentioned are specific to the cloud hence making it as one of the few technologies that allows application developers to suffice the user's needs seamlessly without any disruption.

3.6 Managing the Cloud

Cloud management is aimed at efficiently managing the cloud so as to maintain the QoS. It is one of the prime jobs to be considered. The whole cloud is dependent on the way it is managed. Cloud management can be divided into two parts:

1. Managing the infrastructure of the cloud
2. Managing the cloud application

3.6.1 Managing the Cloud Infrastructure

The infrastructure of the cloud is considered to be the backbone of the cloud. This component is mainly responsible for the QoS factor. If the infrastructure is not properly managed, then the whole cloud can fail and QoS would be adversely affected. The core of cloud management is resource management. Resource management involves several internal tasks such as resource scheduling, provisioning, and load balancing. These tasks are mainly managed by the cloud service provider's core software capabilities such as the cloud OS that is responsible for providing services to the cloud and that internally controls the cloud. A cloud infrastructure is a very complex system that consists of a lot of resources. These resources are usually shared by several users.

Poor resource management may lead to several inefficiencies in terms of performance, functionality, and cost. If a resource is not efficiently managed, the performance of the whole system is affected. Performance is the most important aspect of the cloud, because everything in the cloud is dependent on the SLAs and the SLAs can be satisfied only if performance is good. Similarly, the basic functionality of the cloud should always be provided and considered at any cost. Even if there is a small discrepancy in providing the functionality, the whole purpose of maintaining the cloud is futile. A partially functional cloud would not satisfy the SLAs.

Lastly, the reason for which the cloud was developed was cost. The cost is a very important criterion as far as the business prospects of the cloud are concerned. On the part of the service providers, if they incur less cost for managing the cloud, then they would try to reduce the cost so as to get a strong user base. Hence, a lot of users would use the services, improving their profit margin. Similarly, if the cost of resource management is high, then definitely the cost of accessing the resources would be high and there is never a lossy business from any organization and so the service provider would not bear the cost and hence the users have to pay more. Similarly, this would prove costly for service providers as they have a high chance of losing a wide user base, leading to only a marginal growth in the industry. And, competing with its industry rivals would become a big issue. Hence, efficient management with less cost is required.

At a higher level, other than these three issues, there are few more issues that depend on resource management. These are power consumption and optimization of multiple objectives to further reduce the cost. To accomplish these tasks, there are several approaches followed, namely, consolidation of server and storage workloads. Consolidation would reduce the energy consumption and in some cases would increase the performance of the cloud. According to Margaret Rouse [5], server consolidation by definition is an approach to the efficient usage of computer server resources in order to reduce the total number of servers or server locations that an organization requires.

The previously discussed prospects are mostly suitable for IaaS. Similarly, there are different management methods that are followed for different types of service delivery models. Each of the type has its own way of management. All the management methodologies are based on load fluctuation. Load fluctuation is the point where the workload of the system changes continuously. This is one of the important criteria and issues that should be considered for cloud applications. Load fluctuation can be divided into two types: predictable and unpredictable. Predictable load fluctuations are easy to handle. The cloud can be preconfigured for handling such kind of fluctuations. Whereas unpredictable load fluctuations are difficult to handle, ironically this is one of the reasons why cloud is preferred by several users.

This is as far as cloud management is concerned. Cloud governance is another topic that is closely related to cloud management. Cloud governance is different from cloud management. Governance in general is a term in the corporate world that generally involves the process of creating value to an organization by creating strategic objectives that will lead to the growth of the company and would maintain a certain level of control over the company. Similar to that, here cloud organization is involved.

There are several aspects of cloud governance out of which SLAs are one of the important aspects. SLAs are the set of rules that are defined between the user and cloud service provider that decide upon the QoS factor. If SLAs are not followed, then the defaulter has to pay the penalty. The whole cloud is governed by keeping these SLAs in mind. Cloud governance is discussed in detail in further chapters.

3.6.2 Managing the Cloud Application

Business companies are increasingly looking to move or build their corporate applications on cloud platforms to improve agility or to meet dynamic requirements that exist in the globalization of businesses and responsiveness to market demands. But, this shift or moving the applications to the cloud environment brings new complexities. Applications become more composite and complex, which requires leveraging not only capabilities like storage and database offered by the cloud providers but also third-party SaaS capabilities like e-mail and messaging. So, understanding the availability of an application requires inspecting the infrastructure, the services it consumes, and the upkeep of the application. The composite nature of cloud applications requires visibility into all the services to determine the overall availability and uptime.

Cloud application management is to address these issues and propose solutions to make it possible to have insight into the application that runs in the cloud, as well as implement or enforce enterprise policies like governance and auditing and environment management while the application is deployed in the cloud. These cloud-based monitoring and management services can collect a multitude of events, analyze them, and identify critical information that requires additional remedial actions like adjusting capacity or provisioning new services. Additionally, application management has to be supported with tools and processes required for managing other environments that might coexist, enabling efficient operations.

3.7 Migrating Application to Cloud

Cloud migration encompasses moving one or more enterprise applications and their IT environments from the traditional hosting type to the cloud environment, either public, private, or hybrid. Cloud migration presents an opportunity to significantly reduce costs incurred on applications. This activity comprises, of different phases like evaluation, migration strategy, prototyping, provisioning, and testing.

3.7.1 Phases of Cloud Migration

1. *Evaluation*: Evaluation is carried out for all the components like current infrastructure and application architecture, environment in terms of compute, storage, monitoring, and management, SLAs, operational processes, financial considerations, risk, security, compliance, and licensing needs are identified to build a business case for moving to the cloud.

2. *Migration strategy*: Based on the evaluation, a migration strategy is drawn—a hotplug strategy is used where the applications and their data and interface dependencies are isolated and these applications can be operationalized all at once. A fusion strategy is used where the applications can be partially migrated; but for a portion of it, there are dependencies based on existing licenses, specialized server requirements like mainframes, or extensive interconnections with other applications.

3. *Prototyping*: Migration activity is preceded by a prototyping activity to validate and ensure that a small portion of the applications are tested on the cloud environment with test data setup.

4. *Provisioning*: Premigration optimizations identified are implemented. Cloud servers are provisioned for all the identified environments, necessary platform softwares and applications are deployed, configurations are tuned to match the new environment sizing, and databases and files are replicated. All internal and external integration points are properly configured. Web services, batch jobs, and operation and management software are set up in the new environments.

5. *Testing*: Postmigration tests are conducted to ensure that migration has been successful. Performance and load testing, failure and recovery testing, and scale-out testing are conducted against the expected traffic load and resource utilization levels.

3.7.2 Approaches for Cloud Migration

The following are the four broad approaches for cloud migration that have been adopted effectively by vendors:

1. *Migrate existing applications*: Rebuild or rearchitect some or all the applications, taking advantage of some of the virtualization technologies around to accelerate the work. But, it requires top engineers to develop new functionality. This can be achieved over the course of several releases with the timing determined by customer demand.

2. *Start from scratch*: Rather than cannibalize sales, confuse customers with choice, and tie up engineers trying to rebuild existing application, it may be easier to start again. Many of the R&D decisions will be different now, and with some of the more sophisticated development environments, one can achieve more even with a small focused working team.

3. *Separate company*: One may want to create a whole new company with separate brand, management, R&D, and sales. The investment and internet protocol (IP) may come from the existing company, but many of the conflicts disappear once a new *born in the cloud* company is established. The separate company may even be a subsidiary of the existing company. What is important is that the new company can act, operate, and behave like a cloud-based start-up.

4. *Buy an existing cloud vendor*: For a large established vendor, buying a cloud-based competitor achieves two things. Firstly, it removes a competitor, and secondly, it enables the vendor to hit the ground running in the cloud space. The risk of course is that the innovation, drive, and operational approach of the cloud-based company are destroyed as it is merged into the larger acquirer.

3.8 Summary

Cloud computing has several concepts that must be understood before starting off with the details about the cloud, which include one of the important concepts of cloud architecture. It consists of a basic hierarchical structure with dependencies of components specified. Similarly, anatomy is also important as it describes the basic structure about the cloud, though it does not consider any dependency as in architecture. Further, the cloud network connectivity that forms the core of the cloud model is important. The network is the base using which the cloud works. Similarly, cloud management is one of the important concepts that describe the way in which the cloud is managed, and it has two components: infrastructure management and application management. Both are important as both affect the QoS. Finally, an application should be successfully migrated to a cloud. An application will radiate its complete properties as a cloud only when it has perfectly migrated.

Review Points

- *Cloud architecture*: Cloud architecture consists of a hierarchical set of components that collectively describe the way the cloud works. It is a view of a system (see Section 3.4).

- *Cloud anatomy*: Cloud anatomy is the basic structure of the cloud (see Section 3.5).

- *SLA*: SLAs are a set of agreements that are signed between the user and service providers (see Section 3.8.1).

- *Elasticity*: Elasticity can be defined as the degree to which a system is able to adapt to the workload changes by provisioning and deprovisioning resources in an autonomic manner, such that at each point in time, the available resources match the current demand as closely as possible (see Section 3.7).

- *Multitenancy*: Multitenancy is a property of the cloud by which the software can be shared by different users with full independence (see Section 3.7).

- *Stand-alone application*: A stand-alone application is developed to be run on a single system that does not use a network for its functioning (see Section 3.7).

- *Server consolidation*: Server consolidation by definition is an approach to the efficient usage of computer server resources in order to reduce the total number of servers or server locations that an organization requires (see Section 3.8.1).

Review Questions

1. What is server consolidation?
2. How is cloud anatomy different from cloud architecture?
3. What are the unique properties of cloud applications?
4. What are the two different management classifications?
5. Why are SLAs important?

6. Describe several approaches of cloud migration.

7. What is public cloud access networking?

8. List the phases of cloud migration.

9. What are the drawbacks of a web application?

10. What is elasticity?

11. Explain the pay-as-you-go paradigm.

References

1. The anatomy of cloud computing. Available [Online]: http://www.niallkennedy. com/blog/2009/03/cloud-computing-stack.html. Accessed May 1, 2014.
2. Anatomy of cloud platform. Available [Online]: https://linux.sys-con.com/ node/1120648. Accessed May 2, 2014.
3. Mell, P. and T. Grance. The NIST definition of cloud computing (draft). NIST Special Publication 800.145: 7, 2011.
4. Herbst, N.R., S. Kounev, and R. Reussner. Elasticity in cloud computing: What it is, and what it is not. *Proceedings of the 10th International Conference on Autonomic Computing (ICAC 2013)*, San Jose, CA, 2013.
5. Server consolidation. Available [Online]: http://searchdatacenter.techtarget. com/definition/server-consolidation. Accessed April 28, 2014.

Further Reading

Network considerations in cloud computing. Available [Online]: http://searchcloud computing.techtarget.com/tip/Network-considerations-in-cloud-computing. Accessed May 3, 2014.

4

Cloud Deployment Models

Learning Objectives

The objectives of this book chapter are to

- Introduce the readers to cloud deployment models
- Describe the cloud deployments in detail
- Analyze the advantages and disadvantages of each deployed models
- Discuss the problems related to each deployment model
- Elaborate the deployments models based on properties like SLA and security

Preamble

This chapter broadly discusses the deployment models available in the cloud which are one of the most important concepts related to cloud computing. The deployment models are the different ways in which the cloud computing environment can be set up, that is, the several ways in which the cloud can be deployed. It is important to have an idea about the deployment models because setting up a cloud is the most basic requirement prior to starting any further study about cloud computing. Cloud computing is business oriented, and the popularity of the cloud is credited to its market-oriented nature. In the business perspective, making the correct decision regarding the deployment model is very important. A model should be selected based on the needs, requirements, budget, and security. A wrong decision in the deployment model may affect the organization heavily. Hence, it is very important to know about deployment models. There are many users of the cloud, and each user has different needs. One deployment model will not suite all the cloud users. Based on the cloud setup, the properties of the cloud change. There are four types of deployment models available in the cloud, namely, private, public, community, and hybrid. Each and every type has its own advantages and disadvantages as discussed in the succeeding sections.

4.1 Introduction

Deployment models can be defined as the different ways in which the cloud can be deployed. These models are fully user centric, that is, these depend on users' requirement and convenience. A user selects a model based on his or her requirement and needs. Basically, there are four types of deployment models in the cloud:

1. Private cloud
2. Public cloud
3. Community cloud
4. Hybrid cloud

The classification of the cloud is based on several parameters such as the size of the cloud (number of resources), type of service provider, location, type of users, security, and other issues. The smallest in size is the private cloud (Figure 4.1).

The private cloud is the most basic deployment model that can be deployed by a single organization for its personal use. It is not shared by other organizations, and it is not allowed for public use. The private cloud is to serve the people of an organization. It is usually on premise but can be outsourced also. The next one is the community cloud, which is an extension of the private cloud. Here, the cloud is the same as the private cloud but is shared by several organizations. The community cloud is established for a common cause.

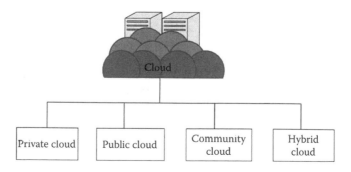

FIGURE 4.1
Cloud deployment models.

The cause can be anything, but usually it leads to mutual benefits among the participating organizations. The next is the public cloud, which is the opposite of the private cloud. This cloud allows access from any place in the world and is open to the public. This cloud is biggest in size among all other deployment models. The public cloud model is one of the most popular deployment models. The public cloud service provider charges the users on an hourly basis and serve the users according to the service-level agreements (SLAs), which are discussed in the succeeding sections. The next one is the hybrid cloud, which is a combination of other deployments. Usually, it consists of the private and public clouds combined. Several properties of the private cloud are used with the properties of the public cloud. This cloud is one of the upcoming cloud models growing in the industry.

All four types of cloud deployments are discussed in detail in subsequent sections.

4.2 Private Cloud

In this section, the private cloud deployment model is discussed. According to the National Institute of Standards and Technology (NIST), private cloud can be defined as the cloud infrastructure that is provisioned for exclusive use by a single organization comprising multiple consumers (e.g., business units). It may be owned, managed, and operated by the organization, a third party, or some combination of them, and it may exist on or off premises [1].

The private cloud in simple terms is the cloud environment created for a single organization. It is usually private to the organization but can be managed by the organization or any other third party. Private cloud can be deployed using Opensource tools such as Openstack [2], Eucalyptus [3].

The private cloud is small in size as compared to other cloud models. Here, the cloud is deployed and maintained by the organizations itself.

4.2.1 Characteristics

Certain characteristics of the private cloud are as follows:

1. *Secure*: The private cloud is secure. This is because usually the private cloud is deployed and managed by the organization itself, and hence there is least chance of data being leaked out of the cloud. In the case of outsourced cloud, the service provider may view the cloud (though governed by SLAs), but there is no other risk from anybody else as all the users belong to the same organization.

2. *Central control*: The organization mostly has full control over the cloud as usually the private cloud is managed by the organization

itself. Thus, when managed by the organization itself, there is no need for the organization to rely on anybody.

3. *Weak SLAs*: Formal SLAs may or may not exist in a private cloud. But if they exist they are weak as it is between the organization and the users of the same organization. Thus, high availability and good service may or may not be available. This depends on the organization that is controlling the cloud.

4.2.2 Suitability

Suitability refers to the instances where this cloud model can be used. It also signifies the most suitable conditions and environment where this cloud model can be used, such as the following:

- The organizations or enterprises that require a separate cloud for their personal or official use.
- The organizations or enterprises that have a sufficient amount of funds as managing and maintaining a cloud is a costly affair.
- The organizations or enterprises that consider data security to be important.
- The organizations that want autonomy and complete control over the cloud.
- The organizations that have a less number of users.
- The organizations that have prebuilt infrastructure for deploying the cloud and are ready for timely maintenance of the cloud for efficient functioning.
- Special care needs to be taken and resources should be available for troubleshooting.

The private cloud platform is not suitable for the following:

- The organizations that have high user base
- The organizations that have financial constraints
- The organizations that do not have prebuilt infrastructure
- The organizations that do not have sufficient manpower to maintain and manage the cloud

According to NIST [4], the private cloud can be classified into several types based on their location and management:

- On-premise private cloud
- Outsourced private cloud

4.2.3 On-Premise Private Cloud

On-premise private cloud is a typical private cloud that is managed by a single organization. Here, the cloud is deployed in organizational premises and is connected to the organizational network. Figure 4.2 describes a private cloud (on premise).

4.2.3.1 Issues

There are several issues associated with private clouds as discussed in the following:

1. *SLA*: SLA plays a very important role in any cloud service deployment model. For any cloud to operate, there must be certain agreements between the user and the service provider. The service provider will agree upon certain terms and conditions regarding the service delivery. These terms and conditions need to be strictly followed; if not, there will be a penalty on the part of the defaulting party. If the service provider fails to provide services as per the SLA, then he has to pay a penalty to the user; this penalty can be in any form, which is termed according to the SLA. These SLAs have different effects on different cloud delivery models. Here in the private cloud, the SLAs are defined between an organization and its users, that is, mostly employees. Usually, these users have broader access rights than the general public cloud users. Similarly in the service provider's side, the service providers are able to efficiently provide the service because of the small user base and mostly efficient network.

2. *Network*: The cloud is totally dependent on the network that is laid out. The network usually consists of a high bandwidth and has a low latency.

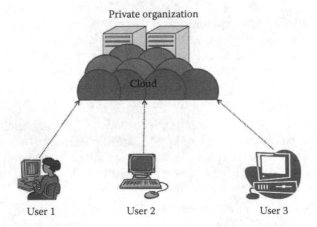

FIGURE 4.2
On-premise private cloud.

This is because the connection is only inside the organization. Network management is easier in this case, and resolving a network issue is easier.

3. *Performance*: The performance of a cloud delivery model primarily depends on the network and resources. Since here the networks are managed internally, the performance can be controlled by the network management team, and mostly this would have good performance as the number of resources is low.

4. *Security and data privacy*: Security and data privacy, though a problem with every type of service model, affect the private cloud the least. As the data of the users are solely managed by the company and most of the data would be related to the organization or company, here there is a lesser chance that the data will be leaked to people outside as there are no users outside the organization. Hence, comparatively, the private cloud is more resistant to attacks than any other cloud type purely because of the type of users and local area network. But, security breaches are possible if an internal user misuses the privileges.

5. *Location*: The private cloud does not have any problems related to the location of data being stored. In a private cloud, the data are internal and are usually stored in the same geographical location where the cloud users, that is, organization, are present (on-premise cloud). If a company has several physical locations, then the cloud is distributed over several places. In this case, there is a possibility that cloud resources have to be accessed using the Internet (by establishing a virtual private network [VPN] or without a VPN).

6. *Cloud management*: Cloud management is a broad area where the entire cloud-related tasks are managed in order to provide seamless services to the customers. This involves several tasks such as resource scheduling, resource provisioning, and resource management. The number of users, the network size, and the amount of resources are some of the important parameters that affect the management of the cloud. Here, the network is small, and the numbers of users and the amount of resources are less.

7. *Multitenancy*: The cloud basically has a multitenant architecture. As multitenant architecture supports multiple tenants with the same physical or software resource, there is a chance of unwanted access of data, and it will have less effect in the private cloud as all the issues will be intraorganizational.

8. *Maintenance*: The cloud is maintained by the organization where the cloud is deployed. The defective resources (drives and processors) are replaced with the good resources. The number of resources is less in the private cloud, so maintenance is comparatively easier.

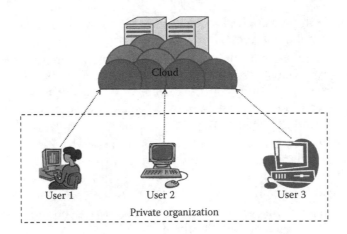

FIGURE 4.3
Outsourced private cloud.

4.2.4 Outsourced Private Cloud

The outsourced private cloud has a cloud outsourced to a third party. A third party manages the whole cloud. Everything is same as usual private cloud except that here the cloud is outsourced. There are several advantages and disadvantages of outsourcing the cloud. The following are the properties that have a significant change due to the outsourced nature of the cloud. All the other aspects are same as on-site private cloud. Figure 4.3 depicts an outsourced private cloud.

4.2.4.1 Issues

The issues that are specific to outsourced private cloud are discussed in the following:

1. *SLA*: The SLA is between the third party and the outsourcing organization. Here, the whole cloud is managed by the third party that will be usually not available on premise. The SLAs are usually followed strictly as it is a third-party organization.
2. *Network*: The cloud is fully deployed at the third-party site. The cloud's internal network is managed by a third party, and the organizations connect to the third party by means of either a dedicated connection or through the Internet. The internal network of the organization is managed by the organization, and it does not come under the purview of the SLA.
3. *Security and privacy*: Security and privacy need to be considered when the cloud is outsourced. Here, the cloud is less secure than the on-site private cloud. The privacy and security of the data mainly depend on the hosting third party as they have the control of the

cloud. But, basically the security threat is from the third party and the internal employee.

4. *Laws and conflicts*: If this cloud is deployed outside the country, then the security laws pertaining to that will apply upon the data and the data are still not fully safe. Usually, private clouds are not deployed outside, but if the off-site location is outside the country's boundary, then several problems may arise.

5. *Location*: The private cloud is usually located off site here. When there is a change of location, the data need to be transmitted through long distances. In few cases, it might be out of the country, which will lead to certain issues regarding the data and its transfer.

6. *Performance*: The performance of the cloud depends on the third party that is outsourcing the cloud.

7. *Maintenance*: The cloud is maintained by a third-party organization where the cloud is deployed. As mentioned, the defective resources (drives and processors) are replaced with the good resources. Here, again the process is less complex compared to the public cloud. The cost of maintenance is a big issue. If an organization owns a cloud, then the cost related to the cloud needs to be borne by the organization and this is usually high.

The deployment of the private cloud into a medium-sized (configuration) machine has now become an easier task. To experience a real cloud, the private cloud can be used. The minimum configuration varies for each type of platforms, but in general, a machine with an 8 GB RAM, 250 GB hard disk, and at least an i7 processor will allow the user to install a private cloud in it. Further, this private (Infrastructure-as-a-Service [IaaS]) cloud can be used to create a virtual machine, and then a user can test these virtual machines. Based on the configuration, the efficiency of the cloud varies. This deployment may not offer a full-fledged private cloud for several users but can be very useful to understand the working of a private cloud.

There are several advantages and disadvantages of a private cloud.

4.2.5 Advantages

- The cloud is small in size and is easy to maintain.
- It provides a high level of security and privacy to the user.
- It is controlled by the organization.

4.2.6 Disadvantages

- For the private cloud, budget is a constraint.
- The private clouds have loose SLAs.

4.3 Public Cloud

According to NIST, the public cloud is the cloud infrastructure that is provisioned for open use by the general public. It may be owned, managed, and operated by a business, academic, or government organization, or some combination of them [1]. It exists on the premises of the cloud provider.

The typical public cloud is depicted in Figure 4.4. Public cloud consists of users from all over the world. A user can simply purchase resources on an hourly basis and work with the resources. There is no need of any prebuilt infrastructure for using the public cloud. These resources are available in the cloud provider's premises. Usually, cloud providers accept all the requests, nd hence, the resources in the service providers' end are considered *infinite* in one aspect. Some of the well-known examples of the public cloud are Amazon AWS [5], Microsoft Azure [6], etc.

4.3.1 Characteristics

1. *Highly scalable*: The public cloud is highly scalable. The resources in the public cloud are large in number and the service providers make sure that all the requests are granted. Hence, the public cloud is considered to be scalable.

2. *Affordable*: The public cloud is offered to the public on a pay-as-you-go basis; hence, the user has to pay only for what he or she is using (usually on a per-hour basis). And, this does not involve any cost related to the deployment.

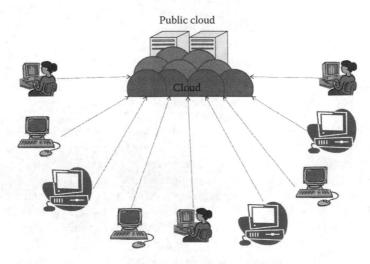

FIGURE 4.4
Public cloud.

3. *Less secure*: The public cloud is less secure out of all the four deployment models. This is because the public cloud is offered by a third party and they have full control over the cloud. Though the SLAs ensure privacy, still there is a high risk of data being leaked.

4. *Highly available*: The public cloud is highly available because anybody from any part of the world can access the public cloud with proper permission, and this is not possible in other models as geographical or other access restrictions might be there.

5. *Stringent SLAs*: SLA is very stringent in the case of the public cloud. As the service provider's business reputation and customer strength are totally dependent on the cloud services, they follow the SLA strictly and violations are avoided. These SLAs are very competitive.

4.3.2 Suitability

There are several occasions and environments where the public cloud is suitable. Thus, the suitability of the public cloud is described. The public cloud can be used whenever the following applies:

- The requirement for resources is large, that is, there is large user base.
- The requirement for resources is varying.
- There is no physical infrastructure available.
- An organization has financial constraints.

The public cloud is not suitable, where the following applies:

- Security is very important.
- Organization expects autonomy.
- Third-party reliability is not preferred.

4.3.3 Issues

Several issues pertaining to the public cloud are as follows:

1. *SLA*: Unlike the private cloud, here the number of users is more and so are the numbers of service agreements. The service provider is answerable to all the users. The users here are diverse. The SLA will cover all the users from all parts of the world. The service provider has to guarantee all the users a fair share without any priority. Having the same SLA for all users is what is usually expected, but it depends on the service provider to have the same SLA for all the users irrespective of the place they are.

2. *Network*: The network plays a major role in the public cloud. Each and every user getting the services of the cloud gets it through

the Internet. The services are accessed through the Internet by all the users, and hence, the service delivery wholly depends on the network. Unlike the private cloud where the organization takes responsibility for the network, here the service provider is not responsible for the network. The service provider is responsible for providing proper service to the customer, and once the services are given from the service provider, it goes on in transit to the user. The user will be charged for even if he or she has problem due to the network. The network usually consists of a high bandwidth and has a low latency. This is because the connection is only inside the organization. Network management is easier in this case.

3. *Performance*: As mentioned, the performance of a cloud delivery model primarily depends on the network and the resources. The service provider has to adequately manage the resources and the network. As the number of users increases, it is a challenging task for the service providers to give good performance.

4. *Multitenancy*: The resources are shared, that is, multiple users share the resources, hence the term multitenant. Due to this property, there is a high risk of data being leaked or a possible unprivileged access.

5. *Location*: The location of the public cloud is an issue. As the public cloud is fragmented and is located in different regions, the access to these clouds involves a lot of data transfers through the Internet. There are several issues related to the location. For example, a user from India might be using the public cloud and he might have to access his personal resources from other countries. This is not good as the data are being stored in some other country.

6. *Security and data privacy*: Security and data privacy are the biggest challenges in the public cloud. As data are stored in different places around the globe, data security is a very big issue. A user storing the data outside his or her country has a risk of the data being viewed by other people as that does not come under the jurisdiction of the user's country. Though this might not always be true, but it may happen.

7. *Laws and conflicts*: The data are stored in different places of the world in different countries. Hence, data centers are bound to laws of the country in which they are located. This creates many conflicts and problems for the service providers and the users.

8. *Cloud management*: Here, the number of users is more, and so the management is difficult. The jobs here are time critical, and as the number of users increases, it becomes more difficult. Inefficient management of resources will lead to resource shortage, and user service might be affected. It has a direct impact on SLA and may cause SLA violation.

9. *Maintenance*: Maintaining the whole cloud is another task. This involves continuous check of the resources, network, and other such parameters for long-lasting efficient delivery of the service. The resource provider has to continuously change the resource components from time to time. The task of maintenance is very crucial in the public cloud. The good the cloud is maintained, the better is the quality of service. Here, the cloud data center is where the maintenance happens; continuously, the disks are replaced from time to time.

The issues discussed earlier will help to understand the public cloud. Before using the public cloud, one has to choose a cloud service provider. One can choose the public cloud based on certain parameters like SLA violations, security, and cost of resources. Thus, a cloud's quality is determined by the SLA violation it does. The less the SLA violation it does, the better the cloud is. This is one way of selecting the public cloud; another way is by cost. If the job for which the resources are used is not time sensitive, then the service provider who offers the least cost is selected.

There following are several advantages and disadvantages of public clouds.

4.3.4 Advantages

- There is no need of establishing infrastructure for setting up a cloud.
- There is no need for maintaining the cloud.
- They are comparatively less costly than other cloud models.
- Strict SLAs are followed.
- There is no limit for the number of users.
- The public cloud is highly scalable.

4.3.5 Disadvantages

- Security is an issue.
- Privacy and organizational autonomy are not possible.

4.4 Community Cloud

According to NIST, the community cloud is the cloud infrastructure that is provisioned for exclusive use by a specific community of consumers from organizations that have shared concerns (e.g., mission, security requirements, policy, and compliance considerations). It may be owned, managed, and operated by one or more of the organizations in the community, a third party, or some combination of them, and it may exist on or off premises [1]. It is a further extension of the private cloud. Here, a private cloud is shared between

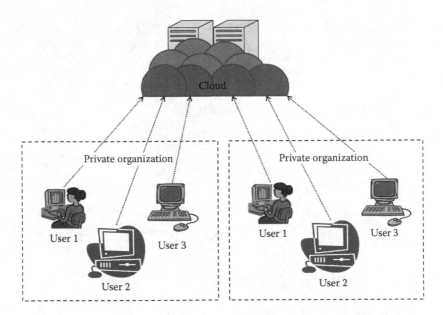

FIGURE 4.5
Community cloud.

several organizations. Either the organizations or a single organization may collectively maintain the cloud.

The main advantage of the public cloud is that the organizations are able to share the resources among themselves based on specific concerns. Thus, here the organizations are able to extract the power of the cloud, which is much bigger than the private cloud, and at the same time, they are able to use it at a usually less cost. The community is formed based on any common cause, but eventually, all the members of the community are benefitted.

This model is very suitable for organizations that cannot afford a private cloud and cannot rely on the public cloud either. Figure 4.5 describes the community cloud.

4.4.1 Characteristics

1. *Collaborative and distributive maintenance*: The community cloud is wholly collaborative, and usually no single party has full control over the whole cloud (in some cases, it may be controlled by one party). This is usually distributive, and hence, better cooperation gives better results. Even though it may be outsourced, collaboration based on purpose always proves to be beneficial.

2. *Partially secure*: Partially secure refers to the property of the community cloud where few organizations share the cloud, so there is a possibility that the data can be leaked from one organization to another, though it is safe from the outside world.

3. *Cost effective*: The community cloud is cost effective as the whole cloud is being shared by several organizations or a community. Usually, not only cost but every other sharable responsibilities are also shared or divided among the groups.

4.4.2 Suitability

This kind of cloud is suitable for organizations that

- Want to establish a private cloud but have financial constraint
- Do not want to complete maintenance responsibility of the cloud
- Want to establish the cloud in order to collaborate with other clouds
- Want to have a collaborative cloud with more security features than the public cloud

This cloud is not suitable for organizations that

- Prefer autonomy and control over the cloud
- Does not want to collaborate with other organizations

There are two types of community cloud deployments:

1. On-premise community cloud
2. Outsourced community cloud

4.4.3 On-Premise Community Cloud

On-premise community cloud consists of the cloud deployed within the premises and is maintained by the organizations themselves.

4.4.3.1 Issues

The issues related to on-site community cloud are as follows:

1. *SLA*: Here, SLA is a little more stringent than the private cloud but is less stringent than the public cloud. As more than one organization is involved, SLA has to be there to have a fair play among the users of the cloud and among the organizations themselves.
2. *Network*: The private cloud can be there in any location as this cloud is being shared by more than one organization. Here, each organization will have a separate network, and they will connect to the cloud. It is the responsibility of each organization to take care of their own network. The service provider is not responsible

for the network issues in the organization. The network is not big and complex as in the public cloud.

3. *Performance*: In this type of deployment, more than one organization coordinate together and provide the cloud service. Thus, it is on the maintenance and management team that the performance depends.

4. *Multitenancy*: There is a moderate risk due to multitenancy. As this cloud is meant for several organizations, the unprivileged access into interorganizational data may lead to several problems.

5. *Location*: The location of the cloud is very important in this case. Usually, the cloud is deployed at any one of the organizations or is maintained off site by any third party. In either case, the organizations have to access the cloud from another location.

6. *Security and privacy*: Security and privacy are issues in the community cloud since several organizations are involved in it. The privacy between the organizations needs to be maintained. As the data are collectively stored, the situation is more like that of a public cloud with less users. The organizations should have complete trust on the service provider, and as all other cloud models, this becomes the bottleneck.

7. *Laws and conflicts*: This applies if organizations are located in different countries. If the organizations are located in the same country, then there is no issue, but if these organizations are located elsewhere, that is, in different countries, then they have to abide by the rules of the country in which the cloud infrastructure is present, thus making the process a bit more complex.

8. *Cloud management*: Cloud management is done by the service provider, here in this case by the organizations collectively. The organizations will have a management team specifically for this cloud and that is responsible for all the cloud management–related operations.

9. *Cloud maintenance*: Cloud maintenance is done by the organizations collectively. The maintenance team collectively maintains all the resources. It is responsible for continuous replacement of resources. In the community cloud, the number of resources is less than the public cloud but usually more than the private cloud.

4.4.4 Outsourced Community Cloud

In the outsourced community cloud, the cloud is outsourced to a third party. The third party is responsible for maintenance and management of the cloud.

4.4.4.1 Issues

The following are some aspects in the community cloud that changed because of the outsourced nature of the community cloud:

1. *SLA*: The SLA is between the group of organizations and the service provider. The SLA here is stringent as it involves a third party. The SLA here is aimed at a fair share of resources among the organizations. The service provider is not responsible for the technical problems within the organization.

2. *Network*: The issues related to the network are same as the on-site community cloud, but here the service provider is outsourced and hence organizations are responsible for their own network and the service provider is responsible for the cloud network.

3. *Performance*: The performance totally depends on the outsourced service provider. The service provider is responsible for efficient services, except for the network issue in the client side.

4. *Security and privacy*: As discussed earlier, there are security and privacy issues as several organizations are involved in it, but in addition to that, the involvement of a third party as a service provider will create much more issues as the organizations have to completely rely on the third party.

5. *Laws and conflicts*: In addition to the issues related to laws due to organizations' location, there is a major issue associated with the location of the cloud service provider. If the service provider is outside the country, then there is conflict related to data laws in that country.

6. *Cloud management and maintenance*: Cloud management and maintenance are done by the service provider. The complexity of managing and maintenance increases with the number of organizations in the community. But, this is less complex than the public cloud.

7. The community cloud as said is an extension of the private cloud. The issues discussed earlier would be more or less the same as the issues related to the private cloud with a very few differences. The community cloud would prove to be successful if a group of organizations work cooperatively.

The following describes the several advantages and disadvantages of the community cloud.

4.4.5 Advantages

- It allows establishing a low-cost private cloud.
- It allows collaborative work on the cloud.

- It allows sharing of responsibilities among the organization.
- It has better security than the public cloud.

4.4.6 Disadvantages

- Autonomy of an organization is lost.
- Security features are not as good as the private cloud.
- It is not suitable if there is no collaboration.

4.5 Hybrid Cloud

According to NIST, the hybrid cloud can be defined as the cloud infrastructure that is a composition of two or more distinct cloud infrastructures (private, community, or public) that remain unique entities but are bound together by standardized or proprietary technology that enables data and application portability [1].

The hybrid cloud usually is a combination of both public and private clouds. This is aimed at combining the advantages of private and public clouds. The usual method of using the hybrid cloud is to have a private cloud initially, and then for additional resources, the public cloud is used. There are several advantages of the hybrid cloud. The hybrid cloud can be regarded as a private cloud extended to the public cloud. This aims at utilizing the power of the public cloud by retaining the properties of the private cloud. One of the popular examples for the hybrid cloud is Eucalyptus [7]. Eucalyptus was initially designed for the private cloud and is basically a private cloud, but now it also supports hybrid cloud. Figure 4.6 shows the hybrid cloud. The hybrid cloud can be further extended into a vast area of federated clouds that is discussed in subsequent chapters.

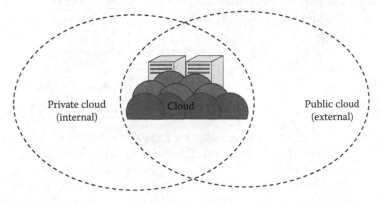

Private cloud (internal) Cloud Public cloud (external)

FIGURE 4.6
Hybrid cloud.

4.5.1 Characteristics

1. *Scalable*: The hybrid cloud is a combination of one or more deployment models. Usually, the private with public cloud gives hybrid cloud. The main reason of having a hybrid cloud is to use the property of a public cloud with a private cloud environment. The public cloud is used whenever needed; hence, as the public cloud is scalable, the hybrid cloud with the help of its public counterpart is also scalable.

2. *Partially secure*: The hybrid cloud usually is a combination of public and private. The private cloud is considered to be secured, but as the hybrid cloud also uses the public cloud, there is high risk of security breach. Thus, it cannot be fully termed as secure but as partially secure.

3. *Stringent SLAs*: As the hybrid cloud involved a public cloud intervention, the SLAs are stringent and might as per the public cloud service provider. But overall, the SLAs are more stringent than the private cloud.

4. *Complex cloud management*: Cloud management is complex and is a difficult task in the hybrid cloud as it involves more than one type of deployment models and also the numbers of users are high.

4.5.2 Suitability

The hybrid cloud environment is suitable for

- Organizations that want the private cloud environment with the scalability of the public cloud
- Organizations that require more security than the public cloud

The hybrid cloud is not suitable for

- Organizations that consider security as a prime objective
- Organizations that will not be able to handle hybrid cloud management

4.5.3 Issues

The cloud can be analyzed in the following aspects:

1. *SLA*: SLA is one of the important aspects of the hybrid cloud as both private and public are involved. There is a right combination of SLAs between the clouds. The private cloud does not have stringent agreements, whereas the public cloud has certain strict rules to be covered. The SLAs to be covered under each purview are clearly defined, and it wholly depends on the service provider (private cloud) to provide efficient services to the customers.

2. *Network*: The network is usually a private network, and whenever there is a necessity, the public cloud is used through the Internet. Unlike the public cloud, here there is a private network also. Thus, a considerable amount of effort is required to maintain the network. The organization takes the responsibility from the network.

3. *Performance*: The hybrid cloud is a special type of cloud in which the private environment is maintained with access to the public cloud whenever required. Thus, here again a feel of an infinite resource is restored. The cloud provider (private cloud) is responsible for providing the cloud.

4. *Multitenancy*: Multitenancy is an issue in the hybrid cloud as it involves the public cloud in addition to the private cloud. Thus, this property can be misused and the breaches will have adverse affects as some parts of the cloud go public.

5. *Location*: Like a private cloud, the location of these clouds can be on premise or off premise and they can be outsourced. They will have all the issues related to the private cloud; in addition to that, issues related to the public cloud will also come into picture whenever there is intermittent access to the public cloud.

6. *Security and privacy*: Whenever the user is provided services using the public cloud, security and privacy become more stringent. As it is the public cloud, the threat of data being lost is high.

7. *Laws and conflicts*: Several laws of other countries come under the purview as the public cloud is involved, and usually these public clouds are situated outside the country's boundaries.

8. *Cloud management*: Here, everything is managed by the private cloud service provider.

9. *Cloud maintenance*: Cloud maintenance is of the same complexity as the private cloud; here, only the resources under the purview of the private cloud need to be maintained. It involves a high cost of maintenance.

The hybrid cloud is one of the fastest growing deployment models, which is now being discussed because of its characteristics as discussed earlier. The issues discussed provide an overview about the difference between the other cloud models and the hybrid cloud model. There is another part of the cloud called as federated cloud that is described in the subsequent chapter.

There are several advantages and disadvantages of the hybrid cloud.

4.5.4 Advantages

- It gives the power of both the private and public clouds.
- It is highly scalable.
- It provides better security than the public cloud.

4.5.5 Disadvantages

- The security features are not as good as the public cloud.
- Managing a hybrid cloud is complex.
- It has stringent SLAs.

4.6 Summary

Cloud computing forms the base for many things in today's world. To start with, the deployment models form the base and need to be known before starting with other aspects of the cloud. These deployment models are based on several properties such as size, location, and complexity. There are four types of deployment models discussed in this chapter. The description of each deployment model with its characteristic and its suitability to different kinds of needs is provided. Each type of deployment model has its own significance. Each deployment model is used in one or other aspects. These deployment models are very important and usually have a great impact on the businesses that are dependent on the cloud. A smart choice of deployment model always proves to be beneficial, avoiding heavy losses. Hence, high importance is given to deployment models.

Review Points

- *Deployment models*: Deployment models can be defined as the different ways in which the cloud can be deployed (see Section 4.1).
- *Private cloud*: Private cloud is the cloud environment created for a single organization (see Section 4.2).
- *Public cloud*: Public cloud is the cloud infrastructure that is provisioned for open use by the general public (see Section 4.3).
- *Hybrid cloud*: Hybrid cloud can be defined as the cloud infrastructure that is a composition of two or more distinct cloud infrastructures (see Section 4.5).
- *Community cloud*: Community cloud is the cloud infrastructure that is provisioned for exclusive use by a specific community of consumers from organizations that have shared concerns (see Section 4.4).
- *SLA*: SLAs are terms and conditions that are negotiated between the service provider and the user (See Section 4.2.3.1).
- *Multitenancy*: Multitenancy is a property of cloud in which multiple users share the same software resource as tenants (see Section 4.2.3.1).

Review Questions

1. Compare and contrast public and private clouds.
2. What is SLA? Are SLAs different for each type of cloud deployment?
3. Analyze the cloud deployment models based on security.
4. How do laws of different countries affect the public cloud model?
5. Differentiate community cloud and hybrid cloud based on their properties.
6. Public cloud is less secure. Justify.
7. What is outsourced community cloud?
8. What are the characteristics of hybrid cloud?
9. What are the advantages of using the community cloud?

References

1. Mell, P. and T. Grance The NIST definition of cloud computing (draft). NIST Special Publication 800.145: 7, 2011.
2. Openstack. Available [Online]: http://www.openstack.org. Accessed April 7, 2014.
3. Eucalyptus: An open source private cloud. Available [Online]: https://www.eucalyptus.com/eucalyptus-cloud/iaas. Accessed April 5, 2014.
4. Badger, L. et al. Cloud computing synopsis and recommendations. NIST Special Publication 800: 146, 2012.
5. Amazon EC2. Available [Online]: http://aws.amazon.com/ec2/. Accessed April 16, 2014.
6. Microsoft Azure. Available [Online]: http://www.azure.microsoft.com/en-us/. Accessed March 20, 2014.
7. Hybrid cloud simplified. Available [Online]: https://www.eucalyptus.com/. Accessed March 12, 2014.

5

Cloud Service Models

Learning Objectives

The main objective of this chapter is to introduce the different service delivery models of cloud computing. After reading this chapter, you will

- Understand the basics of cloud computing stack and cloud service models
- Understand how the Infrastructure as a Service (IaaS) changes computing
- Understand how the Platform as a Service (PaaS) changes the application developer
- Understand how the Software as a Service (SaaS) changes the application delivery
- Understand the characteristics, suitability, and pros and cons of IaaS, PaaS, and SaaS
- Understand the other cloud service models such as Network as a Service (NaaS) and Storage as a Service (STaaS)

Preamble

Cloud computing provides computing resources, development platforms, and applications as a service to the end users. The information technology (IT) industries started subscribing the cloud services instead of buying the products. This chapter gives an insight into the three basic service models of cloud computing, namely, IaaS, PaaS, and SaaS. According to the services provided and subscribed, the responsibilities of the end user and service may vary. This chapter also discusses the responsibilities of the end user and service providers of IaaS, PaaS, and SaaS. Characteristics, suitability, and pros and cons of different cloud service models are also discussed in this chapter along with the summary of popular service providers. At the end, this chapter gives a brief idea about other cloud service models such as NaaS, StaaS, Database as a Service (DBaaS), Security as a Service (SECaaS), and Identity as a Service (IDaaS).

5.1 Introduction

Cloud computing is a model that enables the end users to access the shared pool of resources such as compute, network, storage, database, and application as an on-demand service without the need to buy or own it. The services are provided and managed by the service provider, reducing the management effort from the end user side. The essential characteristics of the cloud include on-demand self-service, broad network access, resource pooling, rapid elasticity, and measured service. The National Institute of Standards and Technology (NIST) defines three basic service models, namely, IaaS, PaaS, and SaaS, as shown in Figure 5.1.

The NIST definition of the three basic service models is given as follows:

1. *IaaS*: The ability given to the infrastructure architects to deploy or run any software on the computing resources provided by the service provider. Here, the underlying infrastructures such as compute, network, and storage are managed by the service provider. Thus, the infrastructure architects are exempted from maintaining the data center or underlying infrastructure. The end users are responsible for managing applications that are running on top of the service

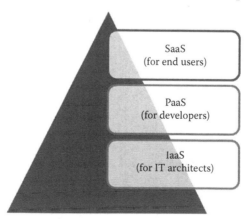

FIGURE 5.1
Basic cloud service models.

provider cloud infrastructure. Generally, the IaaS services are provided from the service provider cloud data center. The end users can access the services from their devices through web command line interface (CLI) or application programming interfaces (APIs) provided by the service providers. Some of the popular IaaS providers include Amazon Web Services (AWS), Google Compute Engine, OpenStack, and Eucalyptus.

2. *PaaS*: The ability given to developers to develop and deploy an application on the development platform provided by the service provider. Thus, the developers are exempted from managing the development platform and underlying infrastructure. Here, the developers are responsible for managing the deployed application and configuring the development environment. Generally, PaaS services are provided by the service provider on an on-premise or dedicated or hosted cloud infrastructure. The developers can access the development platform over the Internet through web CLI, web user interface (UI), and integrated development environments (IDEs). Some of the popular PaaS providers include Google App Engine, Force.com, Red Hat OpenShift, Heroku, and Engine Yard.

3. *SaaS*: The ability given to the end users to access an application over the Internet that is hosted and managed by the service provider. Thus, the end users are exempted from managing or controlling an application, the development platform, and the underlying infrastructure. Generally, SaaS services are hosted in service provider–managed or service provider–hosted cloud infrastructure. The end users can access the services from any thin clients or web browsers. Some of the popular SaaS providers include Saleforce.com, Google Apps, and Microsoft office 365.

The different cloud service models target different audiences. For example, the IaaS model targets the information technology (IT) architects, PaaS targets the developers, and SaaS targets the end users. Based on the services subscribed, the responsibility of the targeted audience may vary as shown in Figure 5.2.

In IaaS, the end users are responsible for maintaining the development platform and the application running on top of the underlying infrastructure. The IaaS providers are responsible for maintaining the underlying hardware as shown in Figure 5.2a. In PaaS, the end users are responsible for managing the application that they have developed. The underlying infrastructure will be maintained by the infrastructure provider as shown in Figure 5.2b. In SaaS, the end user is free from maintaining the infrastructure, development platform, and application that they are using. All the maintenance will be carried out by the SaaS providers as shown Figure 5.2c.

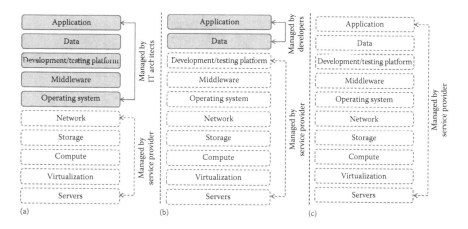

FIGURE 5.2
User and service provider responsibilities of cloud service models: (a) IaaS, (b) PaaS, and (c) SaaS.

The different service models of cloud computing can be deployed and delivered through any one of the cloud deployment models. The NIST defines four different types of cloud deployment models, namely, public cloud, private cloud, community cloud, and hybrid cloud. The public cloud is provided for the general public. The private cloud is used by an organization for its multiple business units. The community cloud is for some group of organization with the same goals. The hybrid cloud is any combination of the public, private, and community clouds. The service delivery of cloud services through different deployment models is shown in Figure 5.3.

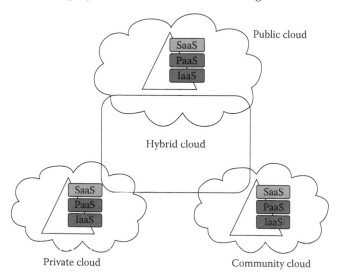

FIGURE 5.3
Deployment and delivery of different cloud service delivery models.

This chapter discusses about the characteristics, suitability, and pros and cons of different cloud service models. Additionally, this chapter gives the summary of popular IaaS, PaaS, and SaaS providers.

5.2 Infrastructure as a Service

IaaS changes the way that the compute, storage, and networking resources are consumed. In traditional data centers, the computing power is consumed by having physical access to the infrastructure. IaaS changes the computing from a physical infrastructure to a virtual infrastructure. IaaS provides virtual computing, storage, and network resources by abstracting the physical resources. Technology *virtualization* is used to provide the virtual resources. All the virtual resources are given to the virtual machines (VMs) that are configured by the service provider. The end users or IT architects will use the infrastructure resources in the form of VMs as shown in Figure 5.4.

The targeted audience of IaaS is the IT architect. The IT architect can design virtual infrastructure, network, load balancers, etc., based on their needs. The IT architects need not maintain the physical servers as it is

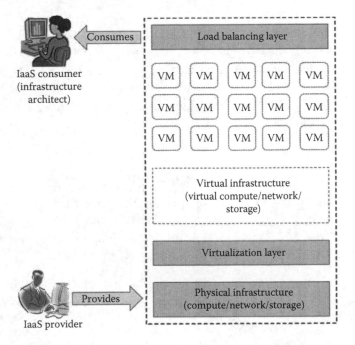

FIGURE 5.4
Overview of IaaS.

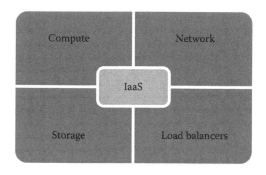

FIGURE 5.5
Services provided by IaaS providers.

maintained by the service providers. The physical infrastructure can be maintained by the service providers themselves. Thus, it eliminates or hides the complexity of maintaining the physical infrastructure from the IT architects. A typical IaaS provider may provide the flowing services as shown in Figure 5.5:

1. *Compute*: Computing as a Service includes virtual central processing units (CPUs) and virtual main memory for the VMs that are provisioned to the end users.
2. *Storage*: STaaS provides back-end storage for the VM images. Some of the IaaS providers also provide the back end for storing files.
3. *Network*: Network as a Service (NaaS) provides virtual networking components such as virtual router, switch, and bridge for the VMs.
4. *Load balancers*: Load Balancing as a Service may provide load balancing capability at the infrastructure layer.

5.2.1 Characteristics of IaaS

IaaS providers offer virtual computing resources to the consumers on a pay-as-you-go basis. IaaS contains the characteristics of cloud computing such as on-demand self-service, broad network access, resource pooling, rapid elasticity, and measured service. Apart from all these, IaaS has its own unique characteristics as follows:

1. *Web access to the resources*: The IaaS model enables the IT users to access infrastructure resources over the Internet. When accessing a huge computing power, the IT user need not get physical access to the servers. Through any web browsers or management console, the users can access the required infrastructure.

2. *Centralized management*: Even though the physical resources are distributed, the management will be from a single place. The resources distributed across different parts can be controlled from any management console. This ensures effective resource management and effective resource utilization.

3. *Elasticity and dynamic scaling*: IaaS provides elastic services where the usage of resources can be increased or decreased according to the requirements. The infrastructure need depends on the load on the application. According to the load, IaaS services can provide the resources. The load on any application is dynamic and IaaS services are capable of proving the required services dynamically.

4. *Shared infrastructure*: IaaS follows a one-to-many delivery model and allows multiple IT users to share the same physical infrastructure. The different IT users will be given different VMs. IaaS ensures high resource utilization.

5. *Preconfigured VMs*: IaaS providers offer preconfigured VMs with operating systems (OSs), network configuration, etc. The IT users can select any kind of VMs of their choice. The IT users are free to configure VMs from scratch. The users can directly start using the VMs as soon as they subscribed to the services.

6. *Metered services*: IaaS allows the IT users to rent the computing resources instead of buying it. The services consumed by the IT user will be measured, and the users will be charged by the IaaS providers based on the amount of usage.

5.2.2 Suitability of IaaS

IaaS reduces the total cost of ownership (TCO) and increases the return on investment (ROI) for start-up companies that cannot invest more in buying infrastructure.

IaaS can be used in the following situations:

1. *Unpredictable spikes in usage*: When there is a significant spike in usage of computing resources, IaaS is the best option for IT industries. When demand is very volatile, we cannot predict the spikes and troughs in terms of demand of the infrastructure. In this situation, we cannot add or remove infrastructure immediately according to the demand in a traditional infrastructure. If there is an unpredictable demand of infrastructure, then it is recommended to use IaaS services.

2. *Limited capital investment*: New start-up companies cannot invest more on buying infrastructure for their business needs. And so by using IaaS, start-up companies can reduce the capital investment on

hardware. IaaS is the suitable option for start-up companies with less capital investment on hardware.

3. *Infrastructure on demand*: Some organizations may require large infrastructure for a short period of time. For this purpose, an organization cannot afford to buy more on-premise resources. Instead, they can rent the required infrastructure for a specific period of time. IaaS best suits the organizations that look for infrastructure on demand or for a short time period.

IaaS helps start-up companies limit its capital expenditure. While it is widely used by start-up companies, there are some situations where IaaS may not be the best option. In following situations, IT users should avoid using the IaaS:

1. *When regulatory compliance does not allow off-premise hosting*: For some companies, its regulation may not allow the application and data to be hosted on third-party off-premise infrastructure.

2. *When usage is minimal*: When the usage is minimal and the available on-premise infrastructure itself is capable of satisfying their needs.

3. *When better performance is required*: Since the IaaS services are accessed through the Internet, sometimes the performance might be not as expected due to network latency.

4. *When there is a need for more control on physical infrastructure*: Some organizations might require physical control over the underlying infrastructure. As the IaaS services are abstracted as virtual resources, it is not possible to have more control on underlying physical infrastructure.

5.2.3 Pros and Cons of IaaS

Being one of the important service models of cloud computing, IaaS provides lot of benefits to the IT users. The following are the benefits provided by IaaS:

1. *Pay-as-you-use model*: The IaaS services are provided to the customers on a pay-per-use basis. This ensures that the customers are required to pay for what they have used. This model eliminates the unnecessary spending on buying hardware.

2. *Reduced TCO*: Since IaaS providers allow the IT users to rent the computing resources, they need not buy physical hardware for running their business. The IT users can rent the IT infrastructure rather than buy it by spending large amount. IaaS reduces the need for buying hardware resources and thus reduces the TCO.

3. *Elastic resources*: IaaS provides resources based on the current needs. IT users can scale up or scale down the resources whenever they want. This dynamic scaling is done automatically using some load balancers. This load balancer transfers the additional resource request to the new server and improves application efficiency.

4. *Better resource utilization*: Resource utilization is the most important criteria to succeed in the IT business. The purchased infrastructure should be utilized properly to increase the ROI. IaaS ensures better resource utilization and provides high ROI for IaaS providers.

5. *Supports Green IT*: In traditional IT infrastructure, dedicated servers are used for different business needs. Since many servers are used, the power consumption will be high. This does not result in Green IT. In IaaS, the need of buying dedicated servers is eliminated as single infrastructure is shared between multiple customers, thus reducing the number of servers to be purchased and hence the power consumption that results in Green IT.

Even though IaaS provides cost-related benefits to small-scale industries, it lacks in providing security to the data. The following are the drawbacks of IaaS:

1. *Security issues*: Since IaaS uses virtualization as the enabling technology, hypervisors play an important role. There are many attacks that target the hypervisors to compromise it. If hypervisors get compromised, then any VMs can be attacked easily. Most of the IaaS providers are not able to provide 100% security to the VMs and the data stored on the VMs.

2. *Interoperability issues*: There are no common standards followed among the different IaaS providers. It is very difficult to migrate any VM from one IaaS provider to the other. Sometimes, the customers might face the vendor lock-in problem.

3. *Performance issues*: IaaS is nothing but the consolidation of available resources from the distributed cloud servers. Here, all the distributed servers are connected over the network. Latency of the network plays an important role in deciding the performance. Because of latency issues, sometimes the VM contains issues with its performance.

5.2.4 Summary of IaaS Providers

There are many public and private IaaS providers in the market who provides infrastructure services to the end users. Table 5.1 provides the summary of popular infrastructure providers.

In the table, the popular IaaS providers are classified based on the license, deployment model, and supported host OS, guest OS, and hypervisors. The end user may choose any IaaS provider that matches their needs. Generally,

TABLE 5.1

Summary of Popular IaaS Providers

Provider	License	Deployment Model	Host OS	Guest OS	Supported Hypervisor(s)
Amazon Web Services	Proprietary	Public	Not available	Red Hat Linux, Windows Server, SuSE Linux, Ubuntu, Fedora, Debian, CentOS, Gentoo Linux, Oracle Linux, and FreeBSD	Xen
Google Compute Engine	Proprietary	Public	Not available	Debian 7 Wheezy, CentOS 6, Red Hat Enterprise Linux, SUSE, Windows Server, CoreOS, FreeBSD, and SELinux	KVM
Microsoft Windows Azure	Proprietary	Public	Not available	Windows Server, CentOS, FreeBSD, openSUSE Linux, and Oracle Enterprise Linux	Windows Azure hypervisor
Eucalyptus	GPLv3	Private and hybrid	Linux	Linux and Windows	Xen, KVM, VMware
Apache CloudStack	Apache 2	Private	Linux	Windows, Linux, and various versions of BSD	KVM, vSphere, XenServer/XCP
OpenNebula	Apache 2	Private, public, and hybrid	CentOS, Debian, and openSUSE	Microsoft Windows and Linux	Xen, KVM, VMware
OpenStack	Apache 2	Private and public	CentOS, Debian, Fedora, RHEL, openSUSE, and Ubuntu	CentOS, Ubuntu, Microsoft Windows, and FreeBSD	libvirt, Hyper-V, VMware, XenServer 6.2, baremetal, docker, Xen, LXC via libvirt

public IaaS consumers need not consider the host OS as it is maintained by the service provider. In managing the private cloud, the users should see the supported host OS. However, most of the private IaaS supports popular guest OS, fully depending on the hypervisor that the IaaS providers are supporting.

5.3 Platform as a Service

PaaS changes the way that the software is developed and deployed. In traditional application development, the application will be developed locally and will be hosted in the central location. In stand-alone application development, the applications will be developed and delivered as executables. Most of the applications developed by traditional development platforms result in a licensing-based software, whereas PaaS changes the application development from local machine to online. PaaS providers provide the development PaaS from the data center. The developers can consume the services over the Internet as shown in Figure 5.6.

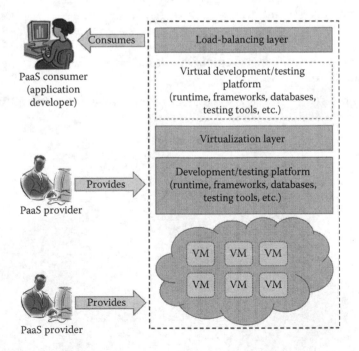

FIGURE 5.6
Overview of PaaS.

PaaS allows the developers to develop their application online and also allows them to deploy immediately on the same platform. PaaS consumers or developers can consume language runtimes, application frameworks, databases, message queues, testing tools, and deployment tools as a service over the Internet. Thus, it reduces the complexity of buying and maintaining different tools for developing an application. Typical PaaS providers may provide programming languages, application frameworks, databases, and testing tools as shown in Figure 5.7. Some of the PaaS providers also provide build tools, deployment tools, and software load balancers as a service:

1. *Programming languages*: PaaS providers provide a wide variety of programming languages for the developers to develop applications. Some of the popular programming languages provided by PaaS vendors are Java, Perl, PHP, Python, Ruby, Scala, Clojure, and Go.

2. *Application frameworks*: PaaS vendors provide application frameworks that simplify the application development. Some of the popular application development frameworks provided by a PaaS provider include Node.js, Rails, Drupal, Joomla, WordPress, Django, EE6, Spring, Play, Sinatra, Rack, and Zend.

3. *Database*: Since every application needs to communicate with the databases, it becomes a must-have tool for every application. PaaS providers are providing databases also with their PaaS platforms. The popular databases provided by the popular PaaS vendors are ClearDB, PostgreSQL, Cloudant, Membase, MongoDB, and Redis.

4. *Other tools*: PaaS providers provide all the tools that are required to develop, test, and deploy an application.

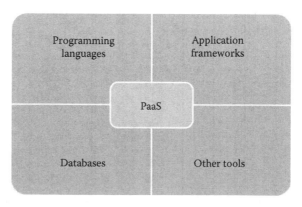

FIGURE 5.7
Services provided by PaaS providers.

5.3.1 Characteristics of PaaS

PaaS development platforms are different from the traditional application development platforms. The following are the essential characteristics that make PaaS unique from traditional development platforms:

1. *All in one*: Most of the PaaS providers offer services to develop, test, deploy, host, and maintain applications in the same IDE. Additionally, many service providers provide all the programming languages, frameworks, databases, and other development-related services that make developers choose from a wide variety of development platforms.

2. *Web access to the development platform*: A typical development platform uses any IDEs for developing applications. Typically, the IDE will be installed in the developer's machines. But, PaaS provides web access to the development platform. Using web UI, any developer can get access to the development platform. The web-based UI helps the developers create, modify, test, and deploy different applications on the same platform.

3. *Offline access*: A developer may not be able to connect to the Internet for a whole day to access the PaaS services. When there is no Internet connectivity, the developers should be allowed to work offline. To enable offline development, some of the PaaS providers allow the developer to synchronize their local IDE with the PaaS services. The developers can develop an application locally and deploy it online whenever they are connected to the Internet.

4. *Built-in scalability*: Scalability is an important requirement for the new-generation web or SaaS applications. It is very difficult to enable the dynamic scalability for any application developed using traditional development platforms. But, PaaS services provide built-in scalability to an application that is developed using any particular PaaS. This ensures that the application is capable of handling varying loads efficiently.

5. *Collaborative platform*: Nowadays, the development team consists of developers who are working from different places. There is a need for a common platform where the developers can collaboratively work together on the same project. Most of the PaaS services provide support for collaborative development. To enable collaboration among developers, most of the PaaS providers provide tools for project planning and communication.

6. *Diverse client tools*: To make the development easier, PaaS providers provide a wide variety of client tools to help the developer. The client tools include CLI, web CLI, web UI, REST API, and IDE. The developers can choose any tools of their choice. These client tools are also capable of handling billing and subscription management.

5.3.2 Suitability of PaaS

Most of the start-up SaaS development companies and independent software vendors (ISVs) widely use PaaS in developing an application. PaaS technology is getting attention from other traditional software development companies also. PaaS is a suitable option for the following situations:

1. *Collaborative development*: To increase the time to market and development efficiency, there is a need for a common place where the development team and other stakeholders of the application can collaborate with each other. Since PaaS services provide a collaborative development environment, it is a suitable option for applications that need collaboration among developers and other third parties to carry out the development process.

2. *Automated testing and deployment*: Automated testing and building of an application are very useful while developing applications at a very short time frame. The automated testing tools reduce the time spent in manual testing tools. Most of the PaaS services offer automated testing and deployment capabilities. The development team needs to concentrate more on development rather than testing and deployment. Thus, PaaS services are the best option where there is a need for automated testing and deployment of the applications.

3. *Time to market*: The PaaS services follow the iterative and incremental development methodologies that ensure that the application is in the market as per the time frame given. For example, the PaaS services are the best option for application development that uses agile development methodologies. If the software vendor wants their application to be in the market as soon as possible, then the PaaS services are the best option for the development.

PaaS is used widely to accelerate the application development process to ensure the time to market. Most of the start-up companies and ISVs started migrating to the PaaS services. Even though it is used widely, there are some situations where PaaS may not be the best option:

1. *Frequent application migration*: The major problem with PaaS services are vendor lock-in. Since there are no common standards followed among PaaS providers, it is very difficult to migrate the application from one PaaS provider to the other.

2. *Customization at the infrastructure level*: PaaS is an abstracted service, and the PaaS users do not have full control over the underlying infrastructure. There are some application development platforms that need some configuration or customization of underlying infrastructure. In these

situations, it is not possible to customize the underlying infrastructure with PaaS. If the application development platform needs any configuration at the hardware level, it is not recommended to go for PaaS.

3. *Flexibility at the platform level*: PaaS provides template-based applications where all the different programming languages, databases, and message queues are predefined. It is an advantage if the application is a generic application.

4. *Integration with on-premise application*: A company might have used PaaS services for some set of applications. For some set of applications, they might have used on-premise platforms. Since many PaaS services use their own proprietary technologies to define the application stack, it may not match with the on-premise application stack. This makes the integration of application hosted in on-premise platform and PaaS platform a difficult job.

5.3.3 Pros and Cons of PaaS

The main advantage of using PaaS is that it hides the complexity of maintaining the platform and underlying infrastructure. This allows the developers to work more on implementing the important functionalities of the application. Apart from this, the PaaS has the following benefits:

1. *Quick development and deployment*: PaaS provides all the required development and testing tools to develop, test, and deploy the software in one place. Most of the PaaS services automate the testing and deployment process as soon as the developer completes the development. This speeds up application development and deployment than traditional development platforms.

2. *Reduces TCO*: The developers need not buy licensed development and testing tools if PaaS services are selected. Most of the traditional development platforms requires high-end infrastructure for its working, which increases the TCO of the application development company. But, PaaS allows the developers to rent the software, development platforms, and testing tools to develop, build, and deploy the application. PaaS does not require high-end infrastructure also to develop the application, thus reducing the TCO of the development company.

3. *Supports agile software development*: Nowadays, most of the new-generation applications are developed using agile methodologies. Many ISVs and SaaS development companies started adopting agile methodologies for application development. PaaS services support agile methodologies that the ISVs and other development companies are looking for.

4. *Different teams can work together*: The traditional development platform does not have extensive support for collaborative development.

PaaS services support developers from different places to work together on the same project. This is possible because of the online common development platform provided by PaaS providers.

5. *Ease of use*: The traditional development platform uses any one of CLI- or IDE-based interfaces for development. Some developers may not be familiar with the interfaces provided by the application development platform. This makes the development job a little bit difficult. But, PaaS provides a wide variety of client tools such as CLI, web CLI, web UI, APIs, and IDEs. The developers are free to choose any client tools of their choice. Especially, the web UI–based PaaS services increase the usability of the development platform for all types of developers.

6. *Less maintenance overhead*: In on-premise applications, the development company or software vendor is responsible for maintaining the underlying hardware. They need to recruit skilled administrators to maintain the servers. This overhead is eliminated by the PaaS services as the underlying infrastructure is maintained by the infrastructure providers. This gives freedom to developers to work on the application development.

7. *Produces scalable applications*: Most of the applications developed using PaaS services are web application or SaaS application. These applications require better scalability on the extra load. For handling extra load, the software vendors need to maintain an additional server. It is very difficult for a new start-up company to provide extra servers based on the additional load. But, PaaS services are providing built-in scalability to the application that is developed using the PaaS platform.

PaaS provides a lot of benefits to developers when compared to the traditional development environment. On the other hand, it contains drawbacks, which are described in the following:

1. *Vendor lock-in*: The major drawback with PaaS providers are vendor lock-in. The main reason for vendor lock-in is lack of standards. There are no common standards followed among the different PaaS providers. The other reason for vendor lock-in is proprietary technologies used by PaaS providers. Most of the PaaS vendors use the proprietary technologies that are not compatible with the other PaaS providers. The vendor lock-in problem of PaaS services does not allow the applications to be migrated from one PaaS provider to the other.

2. *Security issues*: Like in the other cloud services, security is one of the major issues in PaaS services. Since data are stored in off-premise third-party servers, many developers are afraid to go for PaaS

services. Of course, many PaaS providers provide mechanisms to protect the user data, and it is not sufficient to feel the safety of on-premise deployment. When selecting the PaaS provider, the developer should review the regulatory, compliance, and security policies of the PaaS provider with their own security requirements. If not properly reviewed, the developers or users are at the risk of data security breach.

3. *Less flexibility*: PaaS providers do not give much freedom for the developers to define their own application stack. Most of the PaaS providers provide many programming languages, databases, and other development tools. But, it is not extensive and does not satisfy all developer needs. Only some of the PaaS providers allow developers to extend the PaaS tools with the custom or new programming languages. Still most of the PaaS providers do not provide flexibility to the developers.

4. *Depends on Internet connection*: Since the PaaS services are delivered over the Internet, the developers should depend on Internet connectivity for developing the application. Even though some of the providers allow offline access, most of the PaaS providers do not allow offline access. With slow Internet connection, the usability and efficiency of the PaaS platform do not satisfy the developer requirements.

5.3.4 Summary of PaaS Providers

PaaS providers are more in the IT market for public as well as the private clouds. Table 5.2 gives a summary of popular private and public PaaS providers.

5.4 Software as a Service

SaaS changes the way the software is delivered to the customers. In the traditional software model, the software is delivered as a license-based product that needs to be installed in the end user device. Since SaaS is delivered as an on-demand service over the Internet, there is no need to install the software to the end user's devices. SaaS services can be accessed or disconnected at any time based on the end user's needs. SaaS services can be accessed from any lightweight web browsers on any devices such as laptops, tablets, and smartphones. Some of the SaaS services can be accessed from a thin client that does not contain much storage space and cannot run much software like the traditional desktop PCs. The important benefits of using thin clients for accessing the SaaS application are as follows: it is less vulnerable to attack, has a longer life cycle, consumes less power, and is less expensive. A typical

TABLE 5.2

Summary of Popular PaaS Providers

Provider	License	Deployment Model	Supported Languages	Supported Frameworks	Supported Databases	Client Tools
Cloud Foundry	Open source and proprietary	Public	Python, PHP, Java, Groovy, Scala, and Ruby	Spring, Grails, Play, Node.js, Lift, Rails, Sinatra, and Rack	MySQL, PostgreSQL, MongoDB, and Redis	cf. CLI, IDEs, and build tools
Google App Engine	Proprietary	Public	Python, Java, Groovy, JRuby, Scala, Clojure, Go, and PHP	Django, CherryPy, Pyramid, Flask, web2py, and webapp2.	Google Cloud SQL, Datastore, BigTable, and Blobstore	APIs
Heroku	Proprietary	Public	Ruby, Java, Scala, Clojure and Python, PHP, and Perl	Rails, Play, Django, and Node.js.	ClearDB, PostgreSQL, Cloudant, Membase, MongoDB, and Redis	CLI and RESTful API
Microsoft Windows Azure	Proprietary	Public	.Net, PHP, Python, Ruby, and Java	Django, Rails, Drupal, Joomla, WordPress, DotNetNuke, and Node.js.	SQL Azure, MySQL, MongoDB, and CouchDB	RESTful API and IDEs

Red Hat OpenShift Online	Proprietary	Public	Java, Ruby, Python, PHP, and Perl	Node.js, Rails, Drupal, Joomla, WordPress, Django, EE6, Spring, Play, Sinatra, Rack, and Zend.	MySQL, PostgreSQL, and MongoDB	Web UI, APIs, CLI, and IDEs
ActiveState Stackato	Proprietary	Private	Java, Perl, PHP, Python, Ruby, Scala, Clojure, and Go	Spring, Node.js, Drupal, Joomla, WordPress, Django, Rails, and Sinatra.	MySQL, PostgreSQL, MongoDB, and Redis	CLI and IDE
Apprenda	Proprietary	Private	.Net and Java	Most of the frameworks form .Net.	SQL Server	REST APIs
CloudBees	Proprietary	Private	Java, Groovy, and Scala	Spring, JRails, JRuby, and Grails.	MySQL, PostgreSQL, MongoDB, and CouchDB	API, SDK, and IDEs
Cumulogic	Proprietary	Private	Java, PHP, and Python	Spring and Grails.	MySQL, MongoDB, and Couchbase	RESTful API
Gigaspaces Cloudify	Open source	Private	Any programming language specified by recipe	Rails, Play, and others.	MySQL, MongoDB, Couchbase, Cassandra, and others	CLI, web UI, and REST API

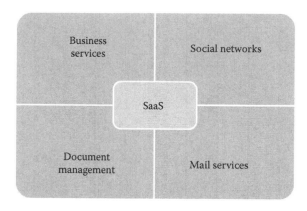

FIGURE 5.8
Services provided by SaaS Providers.

SaaS provider may provide business services, social networks, document management, and mail services as shown in Figure 5.8:

1. *Business services*: Most of the SaaS providers started providing a variety of business services that attract start-up companies. The business SaaS services include ERP, CRM, billing, sales, and human resources.

2. *Social networks*: Since social networking sites are extensively used by the general public, many social networking service providers adopted SaaS for their sustainability. Since the number of users of the social networking sites is increasing exponentially, cloud computing is the perfect match for handling the variable load.

3. *Document management*: Since most of the enterprises extensively use electronic documents, most of the SaaS providers started providing services that are used to create, manage, and track electronic documents.

4. *Mail services*: E-mail services are currently used by many people. The future growth in e-mail usage is unpredictable. To handle the unpredictable number of users and the load on e-mail services, most of the e-mail providers started offering their services as SaaS services.

5.4.1 Characteristics of SaaS

SaaS services are different and give more benefits to end users than the traditional software. The following are the essential characteristics of SaaS services that make it unique from traditional software:

1. *One to many*: SaaS services are delivered as a one-to-many model where a single instance of the application can be shared by multiple tenants or customers.

2. *Web access*: SaaS services provide web access to the software. It allows the end user to access the application from any location if the device is connected to the Internet.

3. *Centralized management*: Since SaaS services are hosted and managed from the central location, management of the SaaS application becomes easier. Normally, the SaaS providers will perform the automatic updates that ensure that each tenant is accessing the most recent version of the application without any user-side updates.

4. *Multidevice support*: SaaS services can be accessed from any end user devices such as desktops, laptops, tablets, smartphones, and thin clients.

5. *Better scalability*: Since most of the SaaS services leverage PaaS and IaaS for its development and deployment, it ensures a better scalability than the traditional software. The dynamic scaling of underlying cloud resources makes SaaS applications work efficiently even with varying loads.

6. *High availability*: SaaS services ensure the 99.99% availability of user data as proper backup and recovery mechanisms are implemented at the back end.

7. *API integration*: SaaS services have the capability of integrating with other software or service through standard APIs.

5.4.2 Suitability of SaaS

SaaS is popular among individuals and start-up companies because of the benefits it provides. Most of the traditional software users are looking for SaaS versions of the software as SaaS has several advantages over traditional applications. SaaS applications are the best option for the following:

1. *On-demand software*: The licensing-based software model requires buying full packaged software and increases the spending on buying software. Some of the occasionally used software does not give any ROI. Because of this, many end users are looking for a software that they can use as and when they needed. If the end users are looking for on-demand software rather than the licensing-based full-term software, then the SaaS model is the best option.

2. *Software for start-up companies*: When using any traditional software, the end user should buy devices with minimum requirements specified by the software vendor. This increases the investment on buying hardware for start-up companies. Since SaaS services do not require high-end infrastructure for accessing, it is a suitable option for start-up companies that can reduce the initial expenditure on buying high-end hardware.

3. *Software compatible with multiple devices*: Some of the applications like word processors or mail services need better accessibility from different devices. The SaaS applications are adaptable with almost all the devices.

4. *Software with varying loads*: We cannot predict the load on popular applications such as social networking sites. The user may connect or disconnect from applications anytime. It is very difficult to handle varying loads with the traditional infrastructure. With the dynamic scaling capabilities, SaaS applications can handle varying loads efficiently without disrupting the normal behavior of the application.

Most of the traditional software vendors moved to SaaS business as it is an emerging software delivery model that attracts end users. But still many traditional applications do not have its SaaS versions. This implies that SaaS applications may not be the best option for all types of software. The SaaS delivery model is not the best option for the applications mentioned in the following:

1. *Real-time applications*: Since SaaS applications depend on Internet connectivity, it may not work better with low Internet speed. If data are stored far away from the end user, the latency issues may delay the data retrieval timings. Real-time applications require fast processing of data that may not be possible with the SaaS applications because of the dependency on high-speed Internet connectivity and latency issues.

2. *Applications with confidential data*: Data security, data governance, and data compliance are always issues with SaaS applications. Since data are stored with third-party service providers, there is no surety that our data will be safe. If the stored confidential data get lost, it will make a serious loss to the organization. It is not recommended to go for SaaS for applications that handle confidential data.

3. *Better on-premise application*: Some of the on-premise applications might fulfill all the requirements of the organization. In such situations, migrating to the SaaS model may not be the best option.

5.4.3 Pros and Cons of SaaS

SaaS applications are used by a wide range of individuals and start-up industries for its cost-related benefits. Apart from the cost-related benefits, SaaS services provide the following benefits:

1. *No client-side installation*: SaaS services do not require client-side installation of the software. The end users can access the services directly from the service provider data center without any installation. There is no need of high-end hardware to consume SaaS

services. It can be accessed from thin clients or any handheld devices, thus reducing the initial expenditure on buying high-end hardware.

2. *Cost savings*: Since SaaS services follow the utility-based billing or pay-as-you-go billing, it demands the end users to pay for what they have used. Most of the SaaS providers offer different subscription plans to benefit different customers. Sometimes, the generic SaaS services such as word processors are given for free to the end users.

3. *Less maintenance*: SaaS services eliminate the additional overhead of maintaining the software from the client side. For example, in the traditional software, the end user is responsible for performing bulk updates. But in SaaS, the service provider itself maintains the automatic updates, monitoring, and other maintenance activities of the applications.

4. *Ease of access*: SaaS services can be accessed from any devices if it is connected to the Internet. Accessibility of SaaS services is not restricted to any particular devices. It is adaptable to all the devices as it uses the responsive web UI.

5. *Dynamic scaling*: SaaS services are popularly known for elastic dynamic scaling. It is very difficult for on-premise software to provide dynamic scaling capability as it requires additional hardware. Since the SaaS services leverage elastic resources provided by cloud computing, it can handle any type of varying loads without disrupting the normal behavior of the application.

6. *Disaster recovery*: With proper backup and recovery mechanisms, replicas are maintained for every SaaS services. The replicas are distributed across many servers. If any server fails, the end user can access the SaaS from other servers. It eliminates the problem of single point of failure. It also ensures the high availability of the application.

7. *Multitenancy*: Multitenancy is the ability given to the end users to share a single instance of the application. Multitenancy increases resource utilization from the service provider side.

Even though SaaS services are used by many individuals and start-up industries, the adoption from the large industries is very low. The major problem with SaaS services is security to the data. All companies are worried about the security of their data that are hosted in the service provider data center. The following are the major problems with SaaS services:

1. *Security*: Security is the major concern in migrating to SaaS application. Since the SaaS application is shared between many end users, there is a possibility of data leakage. Here, the data are stored in the service provider data center. We cannot simply trust some third-party service provider to store our company-sensitive and

TABLE 5.3

Summary of Popular SaaS Providers

Provider	Services Provided
Salseforce.com	On-demand CRM solutions
Google Apps	Gmail, Google Calendar, Talk, Docs, and Sites
Microsoft Office 356	Online office suite, software, plus services
NetSuite	ERP, accounting, order management, inventory, CRM, professional services automation (PSA), and e-commerce applications
Concur	Integrated travel and expense management solutions
GoToMeeting	Online meeting, desktop sharing, and video-conferencing software
Constant Contact	E-mail marketing, social-media marketing, online survey, event marketing, digital storefronts, and local deals tools
Workday, Inc.	Human capital management, payroll, and financial management
Oracle CRM	CRM applications
Intacct	Financial management and accounting software solutions

confidential data. The end user should be careful while selecting the SaaS provider to avoid unnecessary data loss.

2. *Connectivity requirements*: SaaS applications require Internet connectivity for accessing it. Sometimes, the end user's Internet connectivity might be very slow. In such situations, the user cannot access the services with ease. The dependency on high-speed Internet connection is a major problem in SaaS applications.

3. *Loss of control*: Since the data are stored in a third-party and off-premise location, the end user does not have any control over the data. The degree of control over the SaaS application and data is lesser than the on-premise application.

5.4.4 Summary of SaaS Providers

There are many SaaS providers who provide SaaS services such as ERP, CRM, billing, document management, and mail services. Table 5.3 gives a summary of popular SaaS vendors in the market.

5.5 Other Cloud Service Models

The basic cloud services such as IaaS, PaaS, and SaaS are widely used by many individual and start-up companies. Now, cloud computing becomes the dominant technology that drives the IT world. Because of the extensive use of basic cloud services, the end users realize the importance and benefits of specific services such as network, storage, and database. The basic cloud service models are the unified models that contain multiple services in it. Now, the end users' expectation changed, and they are expecting the individual services to be offered by service providers. This makes most of the service providers to think about the separate services that meet end user requirements. Many service providers already started offering separate services such as network, desktop, database, and storage on demand as given in the following:

1. *NaaS* is an ability given to the end users to access virtual network services that are provided by the service provider. Like other cloud service models, NaaS is also a business model for delivering virtual network services over the Internet on a pay-per-use basis. In on-premise data center, the IT industries spent a lot of money to buy network hardware to manage in-house networks. But, cloud computing changes networking services into a utility-based service. NaaS allows network architects to create virtual networks, virtual network interface cards (NICs), virtual routers, virtual switches, and other networking components. Additionally, it allows the network architect to deploy custom routing protocols and enables the design of efficient in-network services, such as data aggregation, stream processing, and caching. Some of the popular services provided by NaaS include virtual private network (VPN), bandwidth on demand (BoD), and mobile network virtualization.

2. *Desktop as a Service (DEaaS)* is an ability given to the end users to use desktop virtualization without buying and managing their own infrastructure. DEaaS is a pay-per-use cloud service delivery model in which the service provider manages the back-end responsibilities of data storage, backup, security, and upgrades. The end users are responsible for managing their own desktop images, applications, and security. Accessing the virtual desktop provided by the DEaaS provider is device, location, and network independent. DEaaS services are simple to deploy, are highly secure, and produce better experience on almost all devices.

3. *STaaS* is an ability given to the end users to store the data on the storage services provided by the service provider. STaaS allows the end users to access the files at any time from any place. The STaaS provider provides the virtual storage that is abstracted from the physical storage of any cloud data center. STaaS is also a cloud business model that is delivered as a utility. Here, the customers can rent the storage from the STaaS provider. STaaS is commonly used as a backup storage for efficient disaster recovery.

4. *DBaaS* is an ability given to the end users to access the database service without the need to install and maintain it. The service provider

is responsible for installing and maintaining the databases. The end users can directly access the services and can pay according to their usage. DBaaS automates the database administration process. The end users can access the database services through any API or web UIs provided by the service provider. The DBaaS eases the database administration process. Popular examples of DBaaS include SimpleDB, DynamoDB, MongoDB as a Service, GAE datastore, and ScaleDB.

5. *Data as a Service (DaaS)* is an ability given to the end users to access the data that are provided by the service provider over the Internet. DaaS provides data on demand. The data may include text, images, sounds, and videos. DaaS is closely related to other cloud service models such as SaaS and STaaS. DaaS can be easily integrated with SaaS or STaaS for providing the composite service. DaaS is highly used in geography data services and financial data services. The advantages of DaaS include agility, cost effectiveness, and data quality.

6. *SECaaS* is an ability given to the end user to access the security service provided by the service provider on a pay-per-use basis. In SECaaS, the service provider integrates their security services to benefit the end users. Generally, the SECaaS includes authentication, antivirus, antimalware/spyware, intrusion detection, and security event management. The security services provided by the SECaaS providers are typically used for securing the on-premise or in-house infrastructure and applications. Some of the SECaaS providers include Cisco, McAfee, Panda Software, Symantec, Trend Micro, and VeriSign.

7. *IDaaS* is an ability given to the end users to access the authentication infrastructure that is managed and provided by the third-party service provider. The end user of IDaaS is typically an organization or enterprise. Using IDaaS services, any organization can easily manage their employees' identity without any additional overhead. Generally, IDaaS includes directory services, federated services, registration, authentication services, risk and event monitoring, single sign-on services, and identity and profile management.

The different new service models discussed in this section emerged after the introduction of cloud computing. This field still evolves and introduces new service models based on the end user's needs. Many researchers from industry and academia already started introducing their innovative idea to take cloud computing to the next level. Apart from the service models discussed in this chapter, cloud computing researchers are thinking to add more service models. Now, cloud computing moves to the scenario where everything can be given as a service. This can be termed as Everything as a Service (XaaS). In the future, we expect many new service models to achieve the goal of XaaS. XaaS may include Backup as a Service (BaaS), Communication as a Service (CaaS), Hadoop as a Service (HaaS), Disaster Recovery as a Service (DRaaS), Testing as a Service (TaaS), Firewall as a Service (FWaaS), Virtual Private Network as a Service (VPNaaS), Load Balancers as a Service (LBaaS), Message Queue as a Service (MQaaS), and Monitoring as a Service (MaaS).

5.6 Summary

Cloud computing composes of three basic service models and four deployment models. The service models include IaaS, PaaS, and SaaS, and the deployment models include private, public, community, and hybrid clouds. The service models decide the type of services provided by each service models. The deployment models decide the way of delivering the services. Cloud computing eliminates a lot of management overhead at each level. IaaS hides the complexity of maintaining the underlying hardware. PaaS hides the complexity of maintaining the development platform and the hardware. In the same way, SaaS hides the complexity of maintaining the application, the development platform, and the hardware. All the basic cloud service models have an essential characteristic of cloud computing: on-demand self-service, broad network access, resource pooling, rapid elasticity, and measured service. Apart from these characteristics, each service model has its own unique characteristics. The characteristics of IaaS include web access to the resources, centralized management, elasticity and dynamic scaling, shared infrastructure, preconfigured VMs, and metered services. The characteristics of PaaS include all in one, web access to the development platform, offline access, built-in scalability, collaborative platform, and diverse client tools. The characteristics of SaaS include one to many, web access, centralized management, multidevice support, better scalability, high availability, and API integration. Cost-based benefits of cloud services make individuals and start-up industries use cloud computing for their computing needs. Even though cloud services are used by many individuals and start-up industries, adaptability from large enterprises is very low. We cannot use cloud services in all places. Generally, the cloud service can be used in start-up companies where the initial capital investment is very low. Cloud services cannot be used when the application uses more sensitive and confidential data. The general benefits of cloud services are cost savings, elastic and dynamic scaling, and centralized management. The general drawbacks include security issues, interoperability issues, and performance issues. Apart from the basic service models, there are other specific cloud services also provided by some vendors, including NaaS, STaaS, DBaaS, SECaaS, and IDaaS.

Review Points

- *SaaS*: The ability given to the end users to access on-demand software services provided by the SaaS provider over the Internet (see Section 5.1).

- *PaaS*: The ability given to the end users to access on-demand development and deployment platforms provided by the PaaS provider over the Internet (See Section 5.1).

- *IaaS*: The ability given to the end users to access on-demand computing resources provided by the IaaS provider over the Internet (see Section 5.1).

- *NaaS*: The ability given to the end users to access on-demand virtual networking components provided by the NaaS provider over the Internet (see Section 5.5).

- *DEaaS*: The ability given to the end users to access on-demand virtual desktop services provided by the DEaaS provider *over the Internet* (see Section 5.5).

- *STaaS*: The ability given to the end users to access on-demand cloud storage provided by the STaaS provider over the Internet (see Section 5.5).

- *DBaaS*: The ability given to the end users to access the on-demand database services provided by the DBaaS provider over the Internet (see Section 5.5).

- *DaaS*: The ability given to the end users to access on-demand data provided by the DaaS provider over the Internet (see Section 5.5).

- *SECaaS*: The ability given to the end users to access security services provided by the SECaaS provider over the Internet (see Section 5.5).

- *IDaaS*: The ability given to the end users to access the identity management services provided by the IDaaS provider over the Internet (see Section 5.5).

Review Questions

1. Define Infrastructure as a Service (IaaS).
2. Define Platform as a Service (PaaS).
3. Define Software as a Service (SaaS).
4. Write short notes on end user and service provider responsibilities of cloud service models with a suitable diagram.
5. Write short notes on the deployment and delivery of cloud service models with a neat diagram.

6. Explain in detail about the overview of IaaS, PaaS, and SaaS with suitable diagrams.

7. Write short notes on the characteristics of IaaS, PaaS, and SaaS.

8. Explain the suitability of different cloud service models.

9. Write short notes on pros and cons of IaaS, PaaS, and SaaS.

10. Write short notes on cloud service models that emerged after the introduction of cloud computing.

Further Reading

Costa, P., M. Migliavacca, P. Pietzuch, and A. L. Wolf. NaaS: Network-as-a-service in the cloud. Hot-ICE, 2012.

Kepes, B. Understanding the cloud computing stack: SaaS, PaaS, IaaS, 2013. Available [Online]: http://www.rackspace.com/knowledge_center/whitepaper/understanding-the-cloud-computing-stack-saas-paas-iaas. Accessed February 18, 2013.

Liu, F., J. Tong, J. Mao, R. B. Bohn, J. V. Messina, M. L. Badger, and D. M. Leaf. NIST cloud computing reference architecture. NIST Special Publication 500-292, 2011. Available [Online]: http://www.nist.gov/customcf/get_pdf.cfm?pub_id=909505. Accessed September 3, 2013.

Orlando, D. Convert cloud computing service models, Part 1: Infrastructure as a service. Technical article. IBM developerWorks, 2011a.

Orlando, D. Convert cloud computing service models, Part 2: Platform as a service. Technical article. IBM developerWorks, 2011b.

Orlando, D. Convert cloud computing service models, Part 3: Software as a service. Technical article. IBM developerWorks, 2011c.

Mell, P. and T. Grance. The NIST definition of cloud computing. NIST Special Publication 800-145, 2011. Available [Online]: http://csrc.nist.gov/publications/nistpubs/800-145/SP800-145.pdf. Accessed September 3, 2013.

Cloud Computing Identity as a Service (IDaaS). Available [Online]: http://www.tutorialspoint.com/cloud_computing/cloud_computing_identity_as_a_service.htm. Accessed September 3, 2013.

Cloud Computing Infrastructure as a Service (IaaS). Available [Online]: http://www.tutorialspoint.com/cloud_computing/cloud_computing_infrastructure_as_a_service.htm. Accessed April 1, 2014.

Cloud Computing Network as a Service (NaaS). Available [Online]: http://www.tutorialspoint.com/cloud_computing/cloud_computing_network_as_a_service.htm. Accessed March 4, 2014.

Cloud Computing Platform as a Service (PaaS). Available [Online]: http://www.tutorialspoint.com/cloud_computing/cloud_computing_platform_as_a_service.htm. Accessed February 27, 2014.

Cloud Computing Software as a Service (SaaS). Available [Online]: http://www.tutorialspoint.com/cloud_computing/cloud_computing_software_as_a_service.htm. Accessed March 20, 2014.

6

Technological Drivers for Cloud Computing

Learning Objectives

The major learning objectives of this chapter are the following:

- To identify the various technological drivers of cloud computing paradigm
- To analyze each underlying technology in detail in order to understand the characteristic features and advantages
- To familiarize with the latest developments in each of these enabling technologies
- To understand how each of these technological components contributes to the success of cloud computing
- To introduce the readers to various case studies in these areas
- To understand how cloud service providers and cloud service consumers are benefitted from these technological advancements in the cloud scenario

Preamble

Cloud computing is an emerging computing paradigm where various users access the resources and services offered by service providers. From a technological perspective, cloud computing is the culmination of many components that enable the present-day cloud computing paradigm to offer the services efficiently to the consumers. Advancements in each of these technological areas have significantly contributed to the widespread adoption of cloud computing. This chapter focuses on identifying those technological drivers of cloud computing and also discusses each technological component in detail. The recent advancements in each of these technologies are highlighted with their advantages and characteristic features. The readers are expected to get the understanding of each of these technological components, their strengths and weaknesses, and also about the benefits provided by these technological drivers to various stakeholders such as service providers and service consumers.

6.1 Introduction

Cloud computing enables service providers to offer various resources such as infrastructure, platform, and software as services to the requesting users on a pay-as-you-go model. The cloud service consumers (CSCs) are benefitted from the cost reduction in procuring the resources and the quality of service (QoS) this promising computing paradigm offers. Nowadays, more companies and enterprises are entering the cloud scenario either as service providers or as service consumers. There has been a considerable increase in the adoption rate of cloud computing among service providers and users across the globe.

The success of cloud computing can be closely associated with the technological enhancements in various areas such as service-oriented architecture (SOA), virtualization, multicore technology, memory and storage technologies, networking technologies, Web 2.0, and Web 3.0. Also, the advancements in programming models, software development models, pervasive computing, operating systems (OSs), and application environment have contributed to the successful deployment of various clouds.

This chapter focuses on various technological drivers for cloud computing and also shows how the latest advancements in each of these enabling technologies have made an impact on the widespread adoption of cloud computing. This chapter also discusses each technological component in detail. The recent advancements in each of these technologies are highlighted with their advantages and characteristic features. The various benefits provided by these technological drivers to different stakeholders such as service providers and service consumers are emphasized.

6.2 SOA and Cloud

Many people are confused thinking that SOA and cloud computing are the same, or cloud computing is another name for SOA. This is not true. SOA is a flexible set of design principles and standards used for systems development and integration. A properly implemented SOA-based system provides a loosely coupled set of services that can be used by the service consumers for meeting their service requirements within various business domains. Cloud computing is a service delivery model in which shared services and resources are consumed by the users across the Internet just like a public utility on an on-demand basis. Generally, SOA is used by enterprise applications, and cloud computing is used for availing the various Internet-based services. Different companies or service providers may offer various services such as financial services, health-care services, manufacturing services, and HR services. Various users can acquire and leverage the offered services through the Internet. In such an environment, cloud is about services and service composition. Cloud offers various infrastructure components such as central processing unit (CPU), memory, and storage. It also provides various development platforms for developing softwares that offer their programmed features to various cloud consumers through service-oriented application programming interfaces (APIs). The programs running on cloud could be implemented using SOA-related technologies. A cloud user can combine the services offered by a cloud service provider (CSP) with other in-house and public cloud services to create SOA-based composite applications.

6.2.1 SOA and SOC

The service-oriented computing (SOC) paradigm utilizes the services for the rapid and low-cost development of interoperable distributed applications in heterogeneous environments. In this paradigm, services are autonomous and platform-independent entities that can be described, published, discovered, and loosely coupled using various protocols and specifications. SOC makes it possible to create cooperating services that are loosely coupled, and these services can be used to create dynamic business processes and agile applications in heterogeneous computing platforms. SOC uses the services architectural model of SOA as shown in Figure 6.1. This model consists of entities such as

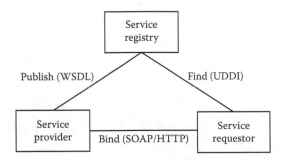

FIGURE 6.1
Services architectural model of SOA.

service provider and service requestor. Service providers publish the details of their services in the service registry using an Extensible Markup Language (XML) called Web Services Description Language (WSDL). Service requestors find the suitable services from the service registry using specifications such as Universal Description, Discovery, and Integration (UDDI). Service providers and service requestors communicate with each other using protocols such as Simple Object Access Protocol (SOAP). SOAP allows a program or service running on one platform to communicate with another program or service running on a different platform, using the Hypertext Transfer Protocol (HTTP) and its XML as the mechanisms for information exchange.

6.2.2 Benefits of SOA

SOA enables mutual data exchange between programs of different vendors without the need for additional programming or changes to the services. The services should be independent, and they should have standard interfaces that can be called to perform their tasks in a standard way. Also, a service need not have prior knowledge of the calling application, and the application does not need to have knowledge about how the tasks are performed by a service. Therefore, the various benefits of SOA are as follows:

1. *Reuse of services*: Various services can be reused by different applications, which results in lower development and maintenance costs. Having reusable services readily available also results in quicker time to market.
2. *Agility*: SOA can bring the architectural agility in an enterprise through the wide use of standards such as web services. This is the ability to change the business processes quickly when needed to support the change in the business activities. This agility aspect helps to deal with system changes using the configuration layer instead of redeveloping the system constantly.
3. *Monitoring*: It helps to monitor the performance of various services to make the required changes.
4. *Extended reach*: In the collaboration between enterprises or in the case of shared processes, it is the ability to get the service of various other processes for completing a particular task.

Hence, SOA can be used as an enabling technology to leverage cloud computing. Cloud computing offers on-demand information technology (IT) resources that could be utilized by extending the SOA outside of the enterprise firewall to the CSPs' domain. This is the process of SOA using cloud computing.

6.2.3 Technologies Used by SOA

There are many technologies and standards used by SOA, which could also be utilized in the cloud computing domain for delivering services efficiently to cloud customers. Some of the standards or protocols are given in the following:

1. *Web services*: Web services can implement an SOA. Web services make functional components or services available for access over the Internet, independent of the platforms and the programming language used. UDDI specification defines a way to publish and discover information about web services.

2. *SOAP*: The SOAP protocol is used to describe the communications protocols.

3. *RPC*: Remote procedure call (RPC) is a protocol that helps a program to request a service from another program located in another computer in a network, without the need to understand network details.

4. *RMI-IIOP*: This denotes the Java remote method invocation (RMI) interface over the Internet Inter-ORB Protocol (IIOP). This protocol is used to deliver Common Object Request Broker Architecture (CORBA) distributed computing capabilities to the Java platform. It supports multiple platforms and programming languages and can be used to execute RPCs on another computer as defined by RMI.

5. *REST*: REpresentational State Transfer (REST) is a stateless architecture that runs over HTTP. It is used for effective interactions between clients and services.

6. *DCOM*: Distributed Component Object Model (DCOM) is a set of Microsoft concepts and program interfaces in which client program can request the services from a server program running on other computers in a network. DCOM is based on the Component Object Model (COM).

7. *WCF (Microsoft implementation of web service forms a part of WCF)*: Windows Communication Foundation (WCF) provides a set of APIs in the .NET Framework for building connected, service-oriented applications.

6.2.4 Similarities and Differences between SOA and Cloud Computing

There are certain common features that SOA and cloud computing share while being different from one another in certain other areas.

6.2.4.1 Similarities

Both cloud computing and SOA share some core principles. First, both rely on the service concept to achieve the objectives. Service is a functionality or a feature offered by one entity and used by another. For example, a service could be retrieving the details of the online bank account of a user. SOA and cloud computing use service delegation in that the required task is delegated either to service provider (in the case of cloud computing) or to other application or business components in the enterprise (in the case of SOA). Service delegation helps the people to use the services without being concerned about the implementation and maintenance details. Services could be shared by multiple applications and users, thereby achieving optimized resource utilization. Second, both cloud computing and SOA promote loose coupling among the components or services, which ensures the minimum dependencies among different parts of the system. This feature reduces the impact that any single change on one part of the system makes on the performance of the overall system. Loose coupling helps the implemented services to be separated and unaware of the underlying technology, topology, life cycle, and organization. The various formats and protocols used in distributed computing, such as XML, WSDL, Interface Description Language (IDL), and Common Data Representation (CDR), help to achieve the encapsulation of technology differences and heterogeneity among the various components used for combining a business solution for solving the computing problems. Various services should be location and technology independent in cloud computing, and SOA can be used for achieving this transparency in the cloud domain.

6.2.4.2 Differences

There are also some differences between the SOA and the cloud computing paradigm. The services in SOA mainly focus on business. Each service in SOA may represent one aspect of the business process. The services could be combined together to provide the required complete business application or business solution. Hence, in this sense, the services are horizontal. At the same time, various services in cloud computing are usually layered such as infrastructure, platform, or software, and the lower layer services support the upper services to deliver applications. Hence, the services in this case are vertical. SOA is used for defining the application architecture. The various components or services of the application are divided based on their roles in the SOA applications. That means the solution for a business problem could be achieved by combining the various abstract services performing the required functions. The services in the SOA can be reused by other applications. Cloud computing is a mechanism for delivering IT services. The various services can be divided or grouped based on their roles such as infrastructure, platform, or software. In this case, for utilizing the cloud services, the consumer does not require a problem before defining the cloud services. The services in this case could also be reused by other applications.

6.2.5 How SOA Meets Cloud Computing

SOA is widely considered to be an enabling technology for cloud computing. In the case of cloud computing, it requires high degree of encapsulation. There should not be any hard dependencies on resource location in order to achieve the true virtualization and elasticity in cloud. Also, threads of execution of various users should be properly isolated in cloud, as any vulnerability will result in the information or data of one user being leaked into another consumer. The web services standards (WS*) used in SOA are also used in the cloud computing domain for solving various issues, such as asynchronous messaging, metadata exchange, and event handling. SOA is an architectural style that is really agnostic of the technology standards adopted in the assembly of composite applications. The service orientation provided by SOA helps in the software design using different pieces of software, each providing separate application functionalities as services to other applications. This feature is independent of any platform, vendor, or technology. Services can be combined by other software applications to provide the complete functionality of a large software application. SOA makes the cooperation of computers connected over a network easy. An arbitrary number of services could be run on a computer, and each service can communicate with any other service in the network without human interaction and also without the need to make any modification to the underlying program itself. Within an SOA, services use defined protocols for transferring and interpreting the messages. WSDL is used to describe the services. The SOAP protocol is used to describe the communications protocols.

SOA is an architecture, and cloud computing is an instance of architecture or an architectural option, not an architecture by itself. When used with cloud computing, SOA helps to deliver IT resources as a service over the Internet, and to mix and match the resources to meet the business requirements. In an enterprise, the database could be hosted with one CSP, process server with another CSP, application development platform with another CSP, and web server with yet another CSP. That means the SOA can be extended to the cloud computing providers to provide a cost-effective solution in such a way that the cloud-based resources and on-premise resources work in tandem. SOA using cloud computing architecture provides the agility in such a way that it could easily be changed to incorporate the business needs since it uses services that are configured through a configuration or process layer.

Cloud and SOA are considered to actually complement each other. SOA is an architectural style for building loosely coupled applications and allows their further composition. It also helps in creating the services that are shared and reused. Cloud computing provides repeatability and standardized easy access to the various shared hardware and software at low cost. In fact, it offers a number of *X as a Service* capabilities. SOA and cloud together provide the required complete services-based solution. Hence, cloud and SOA are required to work together to provide service visibility and service management. Service visibility and governance provide the users the functionality of service discovery within a cloud, and the SOA service management helps in managing the life cycle of services available in cloud. Thus, through the integration of cloud and SOA, cloud can take advantage of the SOA governance approach without the necessity for creating new governance overhead. Having SOA and service orientation in place, the companies or organizations can make adopting cloud services easier and less complex, because a cloud computing environment is also based on services. Both cloud and SOA are focused on delivering services to the business with increased agility, speed, and cost effectiveness.

Multitenancy is the characteristic feature of cloud computing systems. It should be noted that this is a feature possessed by the SOA-based systems. In the multitenant application, a CSP has one instance of a program or the application running on the server, and more than one customer at a time is using the copies of the application. An example is Gmail or Hotmail program. Multitenancy improves the efficiency of the cloud system. In single-tenant applications, only one user at a time is using the application provided by the service provider. An example could be a text editor application used by a single user at a time. In cloud, multitenant applications are preferred as it helps in effective resource utilization. The CSC has to trust the CSP that it will perform its intended tasks or functions without fail.

SOA is considered to be the ideal architecture for cloud computing. Moving to the cloud environment requires having a solid SOA to provide the infrastructure required for successful cloud implementation. Cloud computing is a deployment architecture, and SOA is an architectural approach for how to architect the enterprise IT. Hence, cloud requires the service orientation provided by SOA [1]. With SOA already deployed and executed successfully, taking the full advantage of cloud computing becomes reliable, faster, and more secure. Figure 6.2 shows how SOA can be extended to use the cloud services. The perceived benefits of this integration could be improved collaboration, customer satisfaction, and business growth. By utilizing the SOA for deploying the business capabilities to the cloud environment, businesses can considerably improve the interactions with their business partners and existing customers, thereby increasing their revenues.

6.2.6 CCOA

Cloud computing open architecture (CCOA) [2] is an architecture for the cloud environment that incorporates the SOA. The goals of the CCOA are as follows:

1. To develop an architecture that is reusable and scalable. That means, in future, the architecture should incorporate any further changes without the need for replacing the entire architecture.

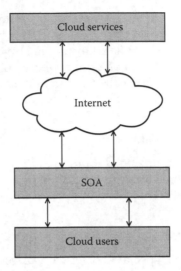

FIGURE 6.2
Convergence of SOA and cloud.

2. To develop a uniform platform for the cloud application development. This will allow the cloud users to switch between the CSPs without the need to make significant changes in the application.

3. To enable the businesses to run efficiently. This goal helps the CSPs to make more money by delivering quality services successfully.

Hence, the CCOA provides the required general guidelines and instructions for the design and development of scalable, reusable, and interoperable cloud applications incorporating SOA principles into them.

6.3 Virtualization

Virtualization is the underlying core technology of cloud computing. It helps in creating a multitenant model for the cloud environment by optimizing the resource usage through sharing (for more details on virtualization you may refer Chapter 7). Benefits of virtualization include the lower costs and extended life of the technology, which has made it a popular option with small- to medium-sized businesses [3]. Using virtualization, the physical infrastructure owned by the service provider is shared among many users, increasing the resource utilization. Virtualization provides efficient resource utilization and increased return on investment (ROI). Ultimately, it results in low capital expenditures (CapEx) and operational expenditures (OpEx).

Some of the benefits of virtualization include better utilization rate of the resources of the service providers, increased ROI for both the service providers and the consumers, and promotes the green IT by reducing energy wastage. Virtualization technology has the drawbacks of the chance of a single point of failure of the software achieving the virtualization and the performance overhead of the entire system due to virtualization.

6.3.1 Approaches in Virtualization

There have been many approaches adopted in the implementation of virtualization technology. Some of the important approaches are discussed in the following subsections.

6.3.1.1 Full Virtualization

Full virtualization uses a special kind of software called a hypervisor. The hypervisor interacts directly with the physical server's hardware resources, such as the CPU and storage space, and acts as a platform for the virtual server's OSs. It helps to keep each virtual server completely independent and unaware of the other virtual servers running on the physical machine. Each guest server or the virtual machine (VM) is able to run its own OS. That means one virtual server could be running on Linux and the other one could be running on Windows. Examples include VMWare ESX and VirtualBox. In the full virtualization, the guest OS is unaware of the underlying hardware infrastructure. That means the guest OS is not aware of the fact that it is running on a virtualized platform and of the feeling that it is running on the real hardware. In this case, the guest OS cannot communicate directly to the underlying physical infrastructure. The OS needs the help of virtualization software hypervisors to communicate with the underlying infrastructure. The advantages of the full virtualization include isolation among the various VMs, isolation between the VMs and the hypervisor, concurrent execution of multiple OSs, and no change required in the guest OS. A disadvantage is that the overall system performance may be affected due to binary translation.

6.3.1.2 Paravirtualization

In this case, VMs do not simulate the underlying hardware, and this uses a special API that a modified guest OS must use. Examples include Xen and VMWare ESX server. In this type of virtualization, partial simulation of the underlying hardware infrastructure is achieved. This is also known as *partial virtualization* or *OS-assisted virtualization*. This virtualization is different from the full virtualization in that, here, the guest OS is aware of the fact that it is running in a virtualized environment. In this case, hypercalls are used for the direct communication between the guest OS and the hypervisor. In paravirtualization, a modified or paravirtualized guest OS is required.

An advantage of this approach is that it improves the overall system performance by eliminating the overhead of binary translation. A disadvantage could be that a modification of the guest OS is required.

6.3.1.3 Hardware-Assisted Virtualization

In this type of virtualization, hardware products supporting the virtualization are used. Hardware vendors like Intel and AMD have developed processors supporting the virtualization through the hardware extension. Intel has released its processor with its virtualization technology VT-x, and AMD have released its processor with its virtualization technology AMD-v to support the virtualization. An advantage of this approach could be that it eliminates the overhead of binary translation and paravirtualization. A disadvantage includes the lack of support from all vendors.

6.3.2 Hypervisor and Its Role

The concept of using VMs increases the resource utilization in a cloud computing environment. Hypervisors are software tools used to create the VMs, and they produce the virtualization of various hardware resources such as CPU, storage, and networking devices. They are also called virtual machine monitor (VMM) or virtualization managers. They help in the virtualization of cloud data centers (DCs). The various hypervisors used are VMware, Xen, Hyper-V, KVM, etc. Hypervisors help to run multiple OSs concurrently on a physical system sharing its hardware. Thus, a hypervisor allows multiple OSs to share a single hardware host. In this case, every OS appears to have the host's processor, memory, and other resources allocated solely to it. However, the hypervisor is actually controlling the host processor and resources and in turn allocates what is needed to each OS. The hypervisor also makes sure that the guest OSs (called VMs) do not interrupt each other. In virtualization technology, hypervisor manages multiple OSs or multiple instances of the same OS on a single physical computer system. Hypervisors are designed to suit a specific processor, and they are also called virtualization managers.

Hypervisors are of mainly two types:

1. *Type 1 hypervisor*: This type of hypervisor runs directly on the host computer's hardware in order to control the hardware resources and also to manage the guest OSs. This is also known as native or bare-metal hypervisors. Examples include VMware ESXi, Citrix XenServer, and Microsoft Hyper-V hypervisor.

2. *Type 2 hypervisor*: This type of hypervisor runs within a formal OS environment. In this type, the hypervisor runs as a distinct second layer while the guest OS runs as a third layer above the hardware. This is also known as the hosted hypervisors. Examples include VMware Workstation and VirtualBox.

6.3.3 Types of Virtualization

Depending on the resources virtualized, the process of virtualization can be classified into the following types.

6.3.3.1 OS Virtualization

In OS virtualization, a desktop's main OS is moved into a virtual environment. The computer that is used by the service consumers remains on their desk, but the OS is hosted on a server elsewhere. Usually, there is one version of the OS on the server, and copies of that individual OS are given to the individual user. Various users can then modify the OS as they wish, without affecting the other users.

6.3.3.2 Server Virtualization

In server virtualization, existing physical servers are moved into a virtual environment, which is then hosted on a physical server. Modern servers can host more than one server simultaneously, which allows the users to reduce the number of servers to be reserved for various purposes. Hence, IT and administrative expenditures are reduced. Server virtualization can use the virtual processors created from the real hardware processor present in the host system. The physical processor can be abstracted into a collection of virtual processors that could be shared by the VMs created.

6.3.3.3 Memory Virtualization

In main memory virtualization, the virtual main memory that is abstracted from the physical memory is allocated to various VMs to meet their memory requirements. The mapping of physical to virtual memory is performed by the hypervisor software. The main memory virtualization support is provided with the modern x86 processors. Also, the main memory consolidation in the virtualized cloud DCs could be performed by the hypervisor by aggregating the free memory segments of various servers to create a virtual memory pool that could be utilized by the VMs.

6.3.3.4 Storage Virtualization

In storage virtualization, multiple physical hard drives are combined into a single virtualized storage environment. To various users, this is simply called cloud storage, and it could be a private storage, such that it is hosted by a company, or a public storage, such that it is hosted outside of a company like DropBox, or a mixed approach of the two. In the case of storage virtualization, physical storage disks are abstracted to a virtual storage media. In cloud DCs, high availability and backup of the user's data are achieved through storage virtualization technology. Modern hypervisors help in achieving storage virtualization. The concept of storage virtualization is implemented in advance storage techniques such as storage area networks (SANs) and network-attached storage (NAS).

6.3.3.5 Network Virtualization

In network virtualization (NV), logical virtual networks are created from the underlying physical network. The physical networking components such as the router, switch, or network interface card could be virtualized by the hypervisor to create logical equivalent components. Multiple virtual networks can be created by using the same physical network components that could be used for various purposes. NV can also be achieved by combining various network components from multiple networks.

6.3.3.6 Application Virtualization

In application virtualization, the single application installed on the central server is virtualized and the various virtualized components of the application will be given to the users requesting the services. In this case, the application is given its own copy of components such as own registry files and global objects that are not shared with others. The virtual environment prevents conflicts in the resource usage. An example is the Java Virtual Machine (JVM). In the cloud computing environment, the CSPs deliver the SaaS model through the application virtualization technology. In the case of application virtualization, the cloud users are not required to install the required applications on their individual systems. They can, in turn, get the virtualized copy of the application, and customize and use it for their own purposes.

6.4 Multicore Technology

In multicore technology, two or more CPUs are working together on the same chip. In this type of architecture, a single physical processor contains the core logic of two or more processors. These processors are packaged into a single integrated circuit (IC). These single ICs are called a die. Multicore technology can also refer to multiple dies packaged together. This technology enables the system to perform more tasks with a greater overall system performance. It also helps in reducing the power consumption and achieving more efficient, simultaneous processing of multiple tasks. Multicore technology can be used in desktops, mobile personal computers (PCs), servers, and workstations. Hence, this technology is used to speed up the processing in a multitenant cloud environment. Multicore architecture has become the recent trend of high-performance processors, and various theoretical and case study results illustrate that multicore architecture is scalable with the number of cores.

6.4.1 Multicore Processors and VM Scalability

In the multicore processor–based system for cloud, state-of-the-art computer architectures are used to allow multiple VMs to scale as long as the cache, memory, bus, and network bandwidth limits are not reached. Thus, in the cloud computing domain, the CPU and memory-intensive virtualized workloads should scale up to the maximum limits imposed by the memory architecture [4].

6.4.2 Multicore Technology and the Parallelism in Cloud

In multicore chips, multiple simpler processors are deployed instead of a single large one, and the parallelism becomes exposed to programmers [5,6]. In the processor design, the development of multicore processors has been a significant recent architectural development. In order to exploit this architectural design fully, the software running on this hardware needs to exhibit concurrent behavior. This fact places greater emphasis on the principles, techniques, and technologies of concurrency. Multicore architecture dominates the server cloud today as it improves the speed and efficiency of processing in cloud. In this case, the main issue is how efficiently the multicore chips are used in the server clouds today in order to gain the real parallelism in terms of performance gain in a multitenant cloud environment through efficient programming.

6.4.3 Case Study

Chip-maker Intel has launched a second-generation family of system on chip (SoC) for microservers [7]. A 64-bit Intel Atom C2000 SoC product is designed for microservers and storage and networking platforms. This is suitable for software-defined networking (SDN). Microservers are tiny, power-efficient machines designed to handle light workloads such as entry-level web hosting. They are suited for small or temporary jobs without provisioning resources from contemporary high-end servers. The Atom C2000 is based on the Silvermont microarchitecture and is aimed at improving performance and energy efficiency. It features up to eight cores, up to 20 W TDP (Thermal Design Power), integrated Ethernet and supports up to 64 GB of memory. It addresses the specialized needs for securing and routing Internet traffic more efficiently. The web-hosting service company 1&1 has tested Atom C2000 as a pilot and is planning to deploy the chip in its entry-level dedicated hosting service. Telecommunications provider Ericsson will also add Atom C2000 SoC to its cloud service platform.

6.5 Memory and Storage Technologies

In the storage domain, it is seen that file-based data growth is faster than block-based data. Also, unstructured data growth is faster than structured data. Most of the organization's data are unstructured, and more than 50% of new storage requirements are consumed by unstructured data such as e-mail, instant messaging, radio frequency identification (RFID) MP3 players, photos, medical imaging, satellite images, and GPS. Hence, the memory or storage solutions used in the cloud environment should support the cloud requirements. The cloud storage has to deal with various kinds of data such as medical images, MP3, photos, 3D high-definition imaging, video streaming, surveillance camera captures, and film animations.

6.5.1 Cloud Storage Requirements

The storage technology or solutions used in the cloud environment should meet the following requirements.

1. *Scalability*: The storage system should support the scalability of the user's data.
2. *High availability*: The degree of availability of the storage solutions deployed in cloud should be very high.
3. *High bandwidth*: The cloud storage system should support the required fast data transfer rate.
4. *Constant performance*: There should not be any performance issues associated with the cloud storage system, and the performance should be consistent throughout the contract period.
5. *Load balancing (LB)*: In order to achieve effective resource usage, the storage systems deployed in cloud should be intelligent enough to support automatic LB of the users' data.

6.5.2 Virtualization Support

In storage virtualization, multiple network storage devices are amalgamated into a single storage unit. This type of virtualization is often used in SAN, which is a high-speed network of shared storage devices, and the SAN technology makes tasks such as archiving, backup, and recovery processes easier and faster. Software applications are used to implement the storage virtualization. Storage virtualization involves the pooling of physical storage from various network storage devices into a single logical storage device, and that is managed from a centralized console. Storage virtualization helps in achieving the easy and efficient backup, archiving, and recovery processes.

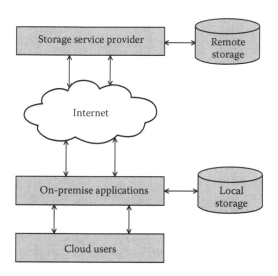

FIGURE 6.3
Storage as a Service.

6.5.3 Storage as a Service (STaaS)

Cloud storage can be internal to the organization or it could be an external storage where the storage is provided by a CSP located outside the organization's DC, which is also known as Storage as a Service (STaaS). STaaS is a cloud business model in which a service provider rents space in its storage infrastructure to various cloud users. Figure 6.3 shows the SaaS cloud model.

The subscribers of the STaaS can have significant cost savings in hardware, maintenance, etc. The STaaS provider agrees in the service-level agreement (SLA) to rent storage space on a cost-per-gigabyte-stored and cost-per-data-transfer basis. STaaS also helps in backup, disaster recovery, business continuity, and availability. Another advantage of STaaS is the capability to access the data stored in cloud from anywhere. Here, the storage is delivered on demand.

6.5.4 Emerging Trends and Technologies in Cloud Storage

The memory and storage technologies have been developing at a rapid pace, and the emerging technologies enable cloud to have a reliable, secure, and scalable storage system. The following are some of the developments in the memory and storage technologies that help the cloud system achieve its efficiency [8,9]:

- Hybrid HDDs with magnetic and flash memory having second-level cache
- Developments in the RAID technology such as RAID 6, RAID triple parity, erasure coding + RAIN

- Converging of block, file, and content data in a single storage subsystem
- Embedded deduplication, primary storage reductions in the storage units
- Usage of object-based storage device (OSD)
- D-RAM SSDs and flash HDDs having the features of improved server utilization, much less energy consumption, being less sensitive to vibration, etc.
- File virtualization or clustered NAS, which supports single namespace to view all files, scaling near linearly by adding nodes, better availability and performance, and LB. These approaches are best suited for seismic processing, video rendering, simulations, auto/aero/electronics design, etc.

The latest models of cloud storage appliances such as Whitewater cloud storage appliance can support up to 14.4 petabytes of logical data. These provide greater capacity, faster speeds, and more replication options. Disk-to-disk backup architectures have become extremely popular in the recent years. Adding the ability to integrate public cloud storage into this architecture offers an immediate return to operations at a disaster recovery site while capturing the cost advantages of very aggressive cloud storage services such as Amazon Glacier.

6.6 Networking Technologies

In cloud, the networking features should support the effective interaction between the CSPs and the CSCs [10].

6.6.1 Network Requirements for Cloud

The various networking requirements in the cloud environment are the following:

1. *Consolidate workloads and provide Infrastructure as a Service (IaaS) to various tenants*: The network technology should enable enterprise workload consolidation, which reduces the management overhead and provides more flexibility and scalability for the management of VMs.
2. *Provide VM connectivity to physical and virtual networks*: The cloud network system should support programmatically managed and extensible features to connect the various VMs to both virtual networks and the physical network. It should also enforce policy enforcement for security, isolation, and service levels.

3. *Ensure connectivity and manage network bandwidth*: The cloud network system should have the load balancing and failover (LBFO) features that allow server bandwidth aggregation and traffic failover.

4. *Speed application and server performance*: The network and OS performance should be improved with high-speed networking features. Various features and technologies such as low-latency technologies, Data Center Bridging, and Data Center Transmission Control Protocol (DCTCP) could be utilized to improve the performance of the cloud networking infrastructure.

6.6.2 Virtualization Support

In the NV network, resources are used through a logical segmentation of a single physical network. NV could be achieved by using software and services to manage the sharing of storage, computing cycles, and the various applications. It considers all the servers and services in the network as a single logical collection of resources that can be accessed without considering the physical components. Hence, NV creates logical, virtual networks that are decoupled or separated from the underlying physical network hardware to ensure that the network is better integrated with the virtual environments. In this case, physical networking devices are simply responsible for the forwarding of the data packets, while the software-controlled virtual network provides the required abstraction that makes it easy to deploy and manage network services and the underlying network resources. Figure 6.4 shows the virtualization support in the networking.

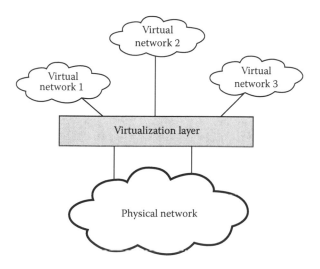

FIGURE 6.4
Virtualization in networking.

NV helps in the efficient utilization of network resources through logical segmentation of a single physical network. This type of virtualization can be used by different organizational units or departments in a company to share the company's physical network as separate logical networks. Here, the total cost of capital expenditures and operating expenses is reduced by sharing the various network resources. In NV, secure logical separation between organizations or groups is maintained. The logical segmentation of the physical network into secure virtual networks could be performed by overlaying virtual private network (VPN) mechanisms such as 802.1x, network admission control (NAC), generic routing encapsulation (GRE) tunnels, virtual routing and forwarding (VRF)-lite, and multiprotocol label switching (MPLS) VPNs onto the existing local area network (LAN).

6.6.3 Usage of Virtual Networks

NV can be used to create virtual networks within a virtualized infrastructure. This helps the NV support the complex resource requirements in multitenant cloud environments. NV can facilitate a virtual network within a virtualized environment that is decoupled from the other network resources. In these virtual environments, NV can separate the data traffic into various zones or containers in order to ensure that the traffic is not mixed with other resources or the transfer of other data.

6.6.4 DCs and VPLS

Truly virtualized DCs should support effective VM migration. In a virtualized DC environment, the VMs can be migrated from one server to another in order to improve the server utilization rate, and also to achieve effective LB. Virtual private LAN service (VPLS) provides the flat network and the long reach that could be used to connect the servers that are geographically apart, thereby achieving VM migration and LB.

6.6.5 SDN

SDN is an approach to networking in which control is decoupled from networking hardware and given to a software application called a controller. In a conventional network, when a data packet arrives at a switch, rules built into the switch's firmware guide the switch in forwarding the data packet. In a software-defined network, a network administrator can control traffic from a centralized control console without having to deal with individual switches. The administrator can change any network switch's rules when necessary such as blocking specific types of packets. This is extremely helpful in cloud computing environment that has a multitenant architecture as it gives the required flexibility and efficiency for the administrator to manage traffic loads. Nowadays, the software-defined network could be created using open standard specifications such as OpenFlow, which allows the network administrators to control routing tables remotely. This could be used in cloud environment for applications that require bandwidth and delay guarantees. Intel has launched Ethernet Switch FM5224 silicon designed to facilitate SDN and to improve compute density and power efficiency.

6.6.6 MPLS

MPLS is a switching mechanism in high-performance telecommunication networks that directs data from one network node to another based on short path labels rather than long network addresses. MPLS sets up a specific path for a given data packet that is identified by a label put in each packet. The various labels identify virtual links or paths between distant nodes rather than end points. This reduces the time needed for a router to look up the address to the next node in data forwarding. MPLS is called multiprotocol because it can be used to encapsulate data packets of different network protocols and technologies such as the Internet Protocol (IP), Asynchronous Transport Mode (ATM), and frame relay network protocols. Thus, MPLS increases the network speed and its manageability.

6.6.7 Other Emerging Networking Trends and Technologies in Cloud

For high bandwidth and ultralow latency in cloud computing, dense wavelength division multiplexing (DWDM) appears to be very promising as a future high-performance WAN transport technology. DWDM is protocol and bit-rate independent, and also it can multiplex multiple optical signals. Researchers are working in new wide area network (WAN) networking technologies such as Lambda networking that promises low-cost, high-capacity circuits in networking. Inside a cloud DC, various network servers, storage systems, network nodes, and other elements are interconnected. Three LAN networking technologies, namely Ethernet, FiberChannel, and InifiBand, are used in the DCs [11]. Ten GB Ethernet (10GBE) equipment and networks are widely used. FiberChannel is suitable for scientific computing and SANs, and InifiBand is almost exclusively deployed in scientific and engineering simulation networks, for example, using clustered servers.

Large cloud DCs support tens of thousands of servers, exabytes of storage, terabits per second of traffic, and tens of thousands of tenants [12]. In a DC, server and storage resources are interconnected with packet switches and routers that provide for the bandwidth and multitenant virtual networking needs. DCs are interconnected across the WAN via routing and transport technologies to provide a pool of resources, known as cloud. High-speed optical interfaces and DWDM optical transport are used to provide for high-capacity transport intra- and inter-DCs.

Private DCs are interconnected through an enterprise-dedicated private network or VPN. Public DCs are connected through the Internet, and they offer multitenant Internet-based services. Virtual private DCs could be built using a common DC infrastructure provided by an IaaS provider. The underlying network for a virtual private DC provides the tenant network isolation and privacy features. Cloud services could be built by interconnecting DCs and utilizing the collective set of resources in these DCs as a resource pool.

Virtualization technology is used to create a VM on a server with dedicated CPU cycles, memory, storage, and input/output (I/O) bandwidth. Multiple VMs for various tenants can be created on the same physical server. A tenant in the cloud environment may also be provided a set of VMs residing on servers distributed throughout a DC or even across various DCs. NV is used as an evolving technology to create intra- and inter-DC networks from the basic virtual local area network (VLAN) and IP routing architecture. This helps to support a large number of tenants and enables networking among their virtualized resources. In the case of large bandwidth requirements such as dealing with large blocks of data and video among DCs, a DC interconnection may utilize an optical transport mechanism. Various network services such as firewalling (FW), service LB, and network address translation (NAT) are provided as part of a tenant virtual network.

VPN technologies are used to carry both Ethernet and IP customer traffic across service provider IP/MPLS networks using tunneling technologies such as IP or MPLS. This helps to achieve the isolation among the customers, and these technologies can be utilized in the cloud domain. In a shared IP/MPLS packet-switched network (PSN), VPLS could be utilized to provide a transparent LAN service. Border Gateway Protocol (BGP) and MPLS IP VPNs [13] are used for providing private customer IP routing over a shared IP/MPLS PSN. Ethernet VPN (EVPN) [14] is an emerging technology aiming to provide Ethernet LAN service over an IP/MPLS PSN. In cloud computing environments, larger switches, higher-speed Ethernet interfaces, and DWDM transport address the bandwidth requirements. Also, evolving packet technologies such as VXLAN as well as Ethernet and IP VPNs are used to create the next-generation DC networking paradigm. Various networking protocols, standards, and technologies are rapidly advancing to meet the requirements of the inter- and intra-DC networks in the cloud domain.

6.7 Web 2.0

Web 2.0 (or Web 2) is the popular term given to the advanced Internet technology and applications that include blogs, wikis, really simple syndication (RSS), and social bookmarking. The two major contributors of Web 2.0 are the technological advances enabled by Ajax and other applications such as

RSS and Eclipse that support the user interaction and their empowerment in dealing with the web. The term was coined by Tim O'Reilly, following a conference dealing with next-generation web concepts and issues held by O'Reilly Media and MediaLive International in 2004. One of the most significant differences between Web 2.0 and the traditional World Wide Web (referred to as Web 1.0) is that Web 2.0 facilitates greater collaboration and information sharing among Internet users, content providers, and enterprises. Hence, in that sense, this can be considered as a migration from the *read-only web* to a *read/write web*.

As an example for this paradigm, multiple-vendor online book outlets such as BookFinder4U allow the users to upload book reviews to the site and also help the users find rare and out-of-print books at a reasonable price. In another example, dynamic encyclopedias such as Wikipedia permit users not only to read the stored information but also to create and edit the contents of the information database in multiple languages. Internet forums such as blogging have become more popular and extensive and have led to the proliferation and sharing of information and views. Also, RSS feeds have been used for the dissemination of news across users and websites.

The main focus of Web 2.0 is to provide the web users the ability to share and distribute information online with other users and sites. It refers to the transition from the static HTML web pages to a more dynamic web for serving web applications to users effectively. A Web 2.0 site such as a social networking site allows its users to interact with each other in a social media dialogue, in contrast to websites where people are restricted to the passive viewing of information [15]. Common examples of Web 2.0 include social networking sites, blogs, wikis, video-sharing sites, and any other hosted services or web applications that allow dynamic sharing of information among users. Also, Web 2.0 technologies can be used as interactive tools to provide feedback on contents or information provided in the web page such as the best practices and recent updates. This feedback will help the service providers to improve the quality of their services and thereby the business values.

Cloud computing is closely related to the SOA of Web 2.0 and virtualization [16] of hardware and software resources. Cloud computing makes it possible to build applications and services that can run/execute utilizing the resources (hardware and software) provided by the service providers, without restricting the application developers or consumers to the resources available on premise.

6.7.1 Characteristics of Web 2.0

In Web 2.0 websites, as already mentioned, instead of merely *reading* the contents from a web page, a user is allowed to *write* or contribute to the content available to everyone in an effective and user-friendly manner. Web 2.0 is also called *network as a platform computing* as it provides software, computing, and storage facilities to the user all through the browser. The major applications of Web 2.0 include social networking sites, self-publishing platforms, tagging, and social bookmarking.

The key features of Web 2.0 include the following:

- Folksonomy
- Rich user experience
- User as a contributor
- User participation
- Dispersion

Folksonomy allows the free classification of information available on the web, which helps the users to collectively classify and find information using approaches such as tagging. Rich user experience is provided because of the dynamic content offered on the web that is responsive to user input in a user-friendly manner. Web 2.0 allows a web user to assume the role of information contributor as information flows in two ways, that is, between site owner and site user by means of evaluation, review, and feedback. This paradigm also facilitates user participation as site users are allowed to add content for others to see (e.g., crowdsourcing). The contributions made by the individual users are available for other users to use and reuse as the web contents. Also, multiple channels are used for content delivery among the users.

Some of the characteristic features of Web 2.0 are as follows [17]:

1. *Blogging*: Blogging allows a user to make a post to a web log or a blog. A blog is a journal, diary, or a personal website that is maintained on the Internet, and it is updated frequently by the user. Blogs increase user interactivity by including features such as comments and links.

2. *Usage of Ajax and other new technologies*: Ajax is a way of developing web applications that combines XHTML and CSS standards–based presentation. It allows the interaction with the web page through the DOM and data interchange with XML and XSLT.

3. *RSS-generated syndication*: RSS is a format for syndicating web content. It allows to *feed* the freshly published web content to the users through the RSS reader/aggregator.

4. *Social bookmarking*: Social bookmarking is a user-defined taxonomy system for storing tags to web contents. The taxonomy is also called *folksonomy,* and the bookmarks are referred to as tags. Instead of storing bookmarks in a folder on the user's computer, tagged pages are stored on the web increasing the accessibility from any computer connected to the Internet.

TABLE 6.1

Differences between Web 1.0 and Web 2.0

Feature	Web 1.0	Web 2.0
Authoring mechanism	Personal websites	Blogging
Information sources	Britannica	Online Wikipedia
Content creation and maintenance	Via CMS	Via wikis
Data storage	Local disk	Online disk
Online advertising	Banners	Google AdSense
Online payment	Bank account	PayPal

5. *Mash-ups*: A mash-up is a web page or an application that can integrate information from two or more sources. Development methodologies using Ajax can be used to create mash-ups. It helps to create a more interactive and participatory web with user-defined contents and services integrated.

6.7.2 Difference between Web 1.0 and Web 2.0

Some of the differences between Web 1.0 and Web 2.0 are shown in Table 6.1.

6.7.3 Applications of Web 2.0

Web 2.0 finds applications in different fields. Some of the applications of Web 2.0 are discussed in the following subsections.

6.7.3.1 Social Media

Social web is an important application of Web 2.0 as it provides a fundamental shift in the way people communicate and share information. The social web offers a number of online tools and platforms that could be used by the users to share their data, perspectives, and opinions among other user communities.

6.7.3.2 Marketing

Web 2.0 offers excellent opportunities for marketing by engaging customers in various stages of the product development cycle. It allows the marketers to collaborate with consumers on various aspects such as product development, service enhancement, and promotion. Collaboration with the business partners and consumers can be improved by the companies by utilizing the tools provided by the Web 2.0 paradigm. Consumer-oriented companies use networks such as Twitter, Yelp, and Facebook as common elements of multichannel promotion of their products. Social networks have become more intuitive and user friendly and can be utilized to disseminate the product information so as to reach the maximum number of prospective product consumers in an efficient manner.

6.7.3.3 Education

Web 2.0 technologies can help the education scenario by providing students and faculty with more opportunities to interact and collaborate with their peers. Effective *knowledge discovery* is possible with the features offered by the Web 2.0 such as greater customization and choice of topics, and less distraction from their peers. By utilizing the tools of Web 2.0, the students get the opportunity to share what they learn with other peers by collaborating with them.

6.7.4 Web 2.0 and Cloud Computing

In Web 2.0, the metadata describing the web content is written in languages such as XML, which can be read and processed by the computers automatically. Various XML-based web protocols such as SOAP, WSDL, and UDDI help to integrate applications developed using different programming languages utilizing heterogeneous computing platforms and OSs. Relying on this capability of data integration and data exchange across heterogeneous applications, new business models of application development, deployment, and delivery over the Internet have been conceptualized and implemented. That means the applications can be hosted on the web and accessed by geographically separated clients over the Internet. Web services are such interoperable applications or services hosted on the web for remote use by multiple clients with heterogeneous platforms, and they can even be discovered dynamically on the fly with no prior knowledge of their existence.

In the business model of cloud computing, the application development infrastructures such as processors, storage, memory, OS, and application development tools and software can be accessed by the clients as services over the Internet in a pay-per-use model. In this model of service delivery, a huge pool of physical resources hosted on the web by the service providers will be shared by multiple clients as and when required. Cloud computing is based on the SOA of Web 2.0 and virtualization [16,18] of hardware and software resources stored hosted by the service providers. Hence, cloud computing is considered as the future of Internet computing because of the advantages offered by this business model such as no capital expenditure, speed of application deployment, shorter time to market, lower cost of operation, and easier maintenance of resources for the clients.

The success of online social networks and other Web 2.0 functionalities encouraged many SaaS applications to offer features that let its users work together, and distribute and share data and information. Cloud computing is a platform that helps individuals and enterprises to access hardware, software, and data resources using the Internet for most of their computing needs [19].

6.8 Web 3.0

The name *Web 3.0* was given by John Markoff of *The New York Times* to this third-generation of the web. The first two generations of the web were called Web 1.0 and Web 2.0 [20]. The three technologies could be briefly described as follows:

Web 1.0: Web 1.0 was the first generation of the Web in which the main focus was building the web, making it accessible, and also commercializing it. The key areas of interest in Web 1.0 included protocols such as HTTP, open standard markup languages such as HTML and XML, Internet access through ISPs, the first web browsers, platforms and tools for web development, web-development software languages such as Java and Javascript, and the commercialization of the web.

Web 2.0: The phrase Web 2.0 was coined by O'Reilly and it refers to the second generation of Internet-based services, such as social networking sites, wikis, and communication tools, that facilitate online collaboration and sharing among various users.

Web 3.0: John Markoff of *The New York Times* coined the term Web 3.0 and it refers to the third generation of Internet-based services that is collectively called *the intelligent web*. Web 3.0 includes services on the Internet that use technologies such as semantic web, natural language search, machine learning, recommendation agents, and artificial intelligence to achieve machine-facilitated understanding of information in order to provide a more productive and intuitive experience to the web users.

Web 2.0 technology allows the use of read/write web, blogs, interactive web applications, rich media, tagging or folksonomy while sharing content, and also social networking sites focusing on communities [21]. At the same time, the Web 3.0 standard uses semantic web technology, drag and drop mash-ups, widgets, user behavior, user engagement, and consolidation of dynamic web contents depending on the interest of the individual users. Web 3.0 uses the *Data Web* technology, which features the data records that are publishable and reusable on the web through query-able formats such as Resource Description Framework (RDF), XML, and microformats. It is an important component facilitating the semantic web, which enables new levels of application interoperability and data integration among various application and services, and also makes data dynamically linkable and accessible in the form of web pages. The complete semantic web stage expands the scope of both structured and unstructured contents through the use of Web Ontology Language (OWL) and RDF semantics.

The Web 3.0 standard also incorporates the latest researches in the field of artificial intelligence. The wide usage of the technology is promptly visible in the case of an application that makes hit-song predictions based on user feedback from music websites hosted by various colleges on the Internet. Web 3.0 achieves the intelligence in an organic fashion through the interaction of the web users. Thus, Web 3.0 makes it possible for an application to think on its own with the data available to make certain decisions, and it also allows to connect one application to another dynamically depending on the context of usage. An example of a typical Web 3.0 application is the one that uses content management systems along with artificial intelligence. These systems are capable of answering the questions posed by the users, because the application is able to think on its own and find the most probable answer, depending on the context, to the query submitted by the user. In this way, Web 3.0 can also be described as a *machine to user* standard in the Internet.

6.8.1 Components of Web 3.0

The term Web 3.0, also known as the *semantic web*, describes sites wherein the computers will be generating raw data on their own without direct user interaction. Web 3.0 is considered as the next logical step in the evolution of the Internet and web technologies. For Web 1.0 and Web 2.0, the Internet is confined within the physical walls of the computer, but as more and more devices such as smartphones, cars, and other household appliances become connected to the web, the Internet will be omnipresent and could be utilized in the most efficient manner. In this case, various devices will be able to exchange data among one another and they will even generate new information from raw data (e.g., a music site, Last.fm, will be able to anticipate the type of music the user likes depending on his previous song selections). Hence, the Internet will be able to perform the user tasks in a faster and more efficient way, such as the case of search engines being able to search for the actual interests of the individual users and not just based on the keyword typed into the search engines.

Web 3.0 embeds intelligence in the entire web domain. It deploys web robots that are smart enough of taking decisions in the absence of any user interference. If Web 2.0 can be called a *read/write* web, Web 3.0 will surely be called a *read/write/execute* web. The two major components forming the basis of Web 3.0 are the following:

1. Semantic web
2. Web services

6.8.1.1 Semantic Web

The semantic web provides the web user a common framework that could be used to share and reuse the data across various applications, enterprises, and community boundaries [22]. The semantic web is a vision of IT that allows the data and information to be readily interpreted by machines, so that the machines are able to take contextual decisions on their own by finding, combining, and acting upon relevant information on the web. The semantic web, as originally envisioned, is a system enabling the machines to *understand* the context and meaning of complex human requests and respond to them appropriately. Also, the semantic web is considered as an integrator of contents or information across different applications and systems.

Web 1.0 represents the first implementation of the web, which, according to Berners-Lee, could be considered as the *read-only web*. This means that the early web allowed the users to search for the required information and read from it. There was very little user interaction or content contribution in this case. The website owners achieved their goal of establishing the online presence and making their information available to all at any time [23].

One of the biggest challenges of presenting information on the web is that web applications are not able to associate the context information to the data, and as a result, they cannot really understand what is relevant and what is not. Through the use of some sort of semantic markup, or data interchange formats, data could be represented in a form that is not only accessible to humans via natural language but also able to be understood and interpreted by software applications. Formatting the data to be understood by software agents is emphasized by the *execute* portion of the *read/write/execute* definition of Web 3.0.

6.8.1.2 Web Services

A web service is a software system that supports computer-to-computer interaction over the Internet. Web services are usually represented as APIs. For example, the popular photography-sharing website Flickr provides a web service that could be utilized by the developers to programmatically interface with Flickr in order to search for images. Currently, thousands of web services are available for users, and they form an important component in the context of Web 3.0. By the combination of semantic markup and web services, the Web 3.0 paradigm promises the potential for applications that can communicate to each other directly and also facilitates broader searches for information through simpler interfaces.

6.8.2 Characteristics of Web 3.0

Web 3.0 could be considered as the third generation of the web and it is enabled by the convergence of several key emerging technology trends as discussed earlier. The major characteristics of this paradigm are the following [24]:

- Ubiquitous connectivity
- Network computing
- Open technologies
- Open identity
- The intelligent web

The Web 3.0 technology enables the continuous connectivity of the user requesting services with various services available through the usage of mobile devices and mobile Internet access with broadband adoption. The Web 3.0 technology enables the Software-as-a-Service (SaaS) business model in cloud computing. Web 3.0 also makes the various web services interoperable by providing open standards. Web 3.0 helps in the creation and usage of open APIs and protocols for service composition and communication. Open data formats and open-source software platforms are supported for the development and use of various applications and services. Web 3.0 also enables the use of open-identity protocols such as OpenID, which helps to port the user account from one service to another effectively.

Web 3.0 supports semantic web technologies (RDF, OWL, SWRL, SPARQL, semantic application platforms, and statement-based datastores such as triplestores, tuplestores, and associative databases) and distributed databases (wide-area distributed database interoperability enabled by semantic web technologies). Also, intelligent applications (using the concepts of natural language processing, machine learning, machine reasoning, and autonomous agents) are created, and their effective communication is made possible. Hence, Web 3.0 helps to achieve a more connected, open, and intelligent web utilizing the aforementioned technologies.

6.8.3 Convergence of Cloud and Web 3.0

The concepts of Web 2.0 and Web 3.0 could be utilized to implement the web as a platform, which forms the basis for the effective delivery and utilization of cloud services [25]. The evolution of Web 3.0 with its enriched capabilities such as personalization, data portability, and user-centric identity provided enough opportunities for enterprises and individuals to generate and use the web contents in a more effective manner, which in turn increased the performance of cloud computing services. Thus, the Web 3.0 paradigm has changed the mode of computing by enabling the service consumers to tap into software and services located in the DCs of service providers rather than confining to the services available on a user's PC. Hence, with the Web 3.0 features, the users could store their data and applications in *cloud* and subscribe to the cloud services as and when needed. Also, the cloud services could be accessed through various devices other than the PC, and the information or the services are made available to various users while moving from one device to another.

With the introduction of Web 2.0, the Internet has been utilized much more than making an online purchase. It has become a gathering place where social communities are formed, personal information is shared, and people dialogue and reconnect with old acquaintances. The launch of social networking sites such as Twitter and Facebook has been a huge success among end users. Because of their ever-growing customer base, these highly interactive social networking sites are used as popular e-marketing tools by business vendors to reach consumers and promote their products through inexpensive and creative ways.

6.8.4 Case Studies in Cloud and Web 3.0

Web 3.0 represents the next generation of web technology that helps achieve an unprecedented level of intelligence and interaction in computing systems and applications. Smarter computing capabilities will be introduced into the web applications in order to perform complex tasks that previously required human interaction for the purposes of understanding and reasoning. With this technology, the various computing tasks could be completed at an enormous scale and efficiency. Today, service providers such as Facebook use these technologies in effective ways. They introduce a variety of innovative capabilities including faster access to the information stored in their DC utilizing the highly intelligent decision-making applications. Some examples are given in the following subsections [26].

6.8.4.1 Connecting Information: Facebook

The Facebook Open Graph is a great example for the scalability feature offered by Web 3.0. Open Graph includes a format for marking up web pages based on the RDF and the semantic web data model so that anybody using a website can incorporate Facebook's markup to define what that site is all about. This goes beyond basic metadata processing because the descriptions enable the web users to find and connect with their friends who share similar interests. The *Like* button provided by Facebook could be considered as a simple manifestation of all these because a single click can offer the analysts an invaluable amount of information that could later be used for further communication with friends and also to make recommendations and discoveries.

6.8.4.2 Search Optimization and Web Commerce: Best Buy

One of the major benefits of Web 3.0 is more relevant search results. In the case of Web commerce, this means that additional information can be incorporated into product descriptions and online ads in order to make them easier for search engines to find. Best Buy is a frontrunner in using this technology to leverage its e-commerce efforts. It is using RDFa (RDF in attributes, which adds a set of XHTML attributes) markup and the GoodRelations vocabulary so that more targeted search results are produced for shoppers looking for various products. So far, some internal company audit measures suggest that this approach has increased the consumer traffic by 30%.

6.8.4.3 Understanding Text: Millward Brown

Web 3.0 is ideally suited for the management and analysis of various documents and information because it enables computing system to quickly process large amounts of text and extract the meaning from them. *Sentiment analysis* could be considered as a good example for this, which involves the measurement of how various customers feel about an organization, as expressed through surveys, blogs, online forums, and social networks. The global research agency Millward Brown, which works with Fortune 500 companies for developing their branding strategies, uses Web 3.0 technologies from OpenAmplify to identify the strategically meaningful information extracted from customer feedback. This information could then be used to drive marketing messages and also to improve public relations efforts, pricing strategies, and service responses of the companies.

6.9 Software Process Models for Cloud

The success or quality of a software project is measured by whether it is developed within time and budget and by its efficiency, usability, dependability, and maintainability [27,28]. The whole development process of software from its conceptualization to operation and retirement is called the software development life cycle (SDLC). SDLC goes through several framework activities like requirements gathering, planning, design, coding, testing, deployment, maintenance, and retirement. These activities are synchronized in accordance to the process model adopted for a particular software development.

There are many process models to choose from, depending on the size of the project, delivery time requirement, and type of the project. For example, the process model selected for the development of an avionic embedded system will be different from the one selected for the development of a web application.

6.9.1 Types of Software Models

There are various software development models or methodologies such as waterfall, V, incremental, RAD, agile, iterative, and spiral. They are discussed in the following.

6.9.1.1 Waterfall Model

This is the most common life cycle model and is also referred to as a linear-sequential life cycle model. In a waterfall model, each phase must be completed in its entirety before the next phase begins. At the end of each phase, a review takes place to determine if the project is on the right path and whether or not to continue the project.

6.9.1.2 V Model

V model means verification and validation model. Just like the waterfall model, the V model is a sequential path of execution of processes. Each phase must be completed before the next phase begins. Testing of the product is planned in parallel with a corresponding phase of development.

6.9.1.3 Incremental Model

The incremental model is an intuitive approach to the waterfall model. Multiple development cycles take place here, making the life cycle a *multi-waterfall* cycle. The cycles are divided into smaller, more easily managed iterations. The iterations pass through the requirements, design, implementation, and testing phases, and during the first iteration, a working version of the software is generated.

6.9.1.4 RAD Model

The rapid application development (RAD) model is a type of incremental model. In the RAD model, the functions or components are generated in parallel, and these generated outcomes are timeboxed, delivered, and then combined to a working prototype. This can quickly give the customer something to operate and to give feedback about the requirements.

6.9.1.5 Agile Model

The agile model is also an incremental model where the software is developed in rapid, incremental cycles. The development results in tiny incremental releases and is based on previously built functionality and is carefully tested to ensure software quality. In time-critical applications, this model is preferred more. Extreme programming (XP) is one popular example of this developmental life cycle model.

6.9.1.6 Iterative Model

It does not start with a full specification of requirements, but it begins by specifying and implementing just a part of the software, which can then be reviewed in order to identify further requirements. This process is then repeated, producing a new version of the software for each cycle of the model.

6.9.1.7 Spiral Model

The spiral model is similar to the incremental model, with more emphasis on the risk analysis. The spiral model has four phases: planning, risk analysis, engineering, and evaluation. In this model, a software project repeatedly passes through these phases in iterations called spirals. The baseline spiral starts at the planning phase, requirements are gathered, and risk is assessed. Each subsequent spiral builds on the baseline spiral.

6.9.2 Agile SDLC for Cloud Computing

In the rapidly changing computing environment in web services and cloud platforms, software development is going to be very challenging [29]. Software development process will involve heterogeneous platforms, distributed web services, and multiple enterprises geographically dispersed all over the world. Existing software process models and framework activities are not adequate unless interaction with cloud providers is included. Requirements gathering phase so far included customers, users, and software engineers. Now, it has to include the cloud providers as well, as they will be providing the computing infrastructure and its maintenance. As only the cloud providers will know the size, architectural details, virtualization strategy, and resource utilization of the infrastructure, they should also be included in the planning and design phases of software development. Coding and testing can be done on the cloud platform, which is a huge benefit as everybody will have easy access to the software being built. This will reduce the cost and time for testing and validation.

In the cloud environment, software developers can use the web services and open-source software freely available from the cloud instead of procuring them. Software developers build software from readily available components rather than writing it all and building a monolithic application. Refactoring of existing application is required to best utilize the cloud infrastructure architecture in a cost-effective way. In the latest hardware technology, the computers are multicore and networked and the software engineers should train themselves in parallel and distributed computing to complement these advances of hardware and network technology. Cloud providers will insist that software should be as modular as possible for occasional migration from one server to another for LB as required by the cloud provider [30].

SDLC is a framework that defines tasks to be performed at each step in the software development process [31]. Cloud computing provides an almost instant access to the software and development environments, by providing multitenancy of the virtualized servers and other IT infrastructures. Specifically, Platform as a Service (PaaS), the development platform environment in the cloud, encourages the use of agile methodologies. Agile and PaaS together add great value to the SDLC processes. They help in reducing costs for enterprises in the long run and help in increasing developer productivity at the same time.

6.9.2.1 Features of Cloud SDLC

SDLC for cloud computing is different from the traditional SDLC in the following ways:

1. *Inclination toward agile methodologies*: Cloud SDLC can utilize methodologies such as agile SDLC. These are designed for iterative approach to development and fast deployment life cycles.
2. *Customizable SDLC framework for different stages*: Cloud computing SDLC must have the capabilities to be customized according to the requirements of the project. In other words, the elasticity and robustness of cloud computing environment can be best utilized if the SDLCs for cloud are customizable.
3. *Installation and configuration guidelines*: SDLC for cloud must provide implementation approach and guidelines for installation and configuration of the cloud depending on its size. The guidelines must ensure that installation and configuration of infrastructure and application environment are completed appropriately for different stages of SDLC including operations and maintenance. These guidelines are the key to differentiating SDLC for cloud from traditional SDLC.

6.9.3 Agile Software Development Process

More than 50% of software projects fail due to various reasons like schedule and budget slippage, non-user-friendly interface of the software, and nonflexibility for maintenance and change of the software [29]. The reason for all these problems is lack of communication and coordination between all the parties involved. Requirement changes of a software are the major cause of increased complexity, schedule, and budget slippage. Incorporating changes at a later stage of SDLC increases cost of the project exponentially. Adding more programmers at a later stage does not solve the schedule problem as increased coordination requirement slows down the project further. It is very important that requirements gathering, planning, and design of the software are done involving all the concerned parties from the beginning of the project.

That is why several agile process models like XP, Scrum, Crystal, and Adaptive have been introduced in the mid-1990s to accommodate continuous changes in requirements during the development of the software. These agile process models have shorter development cycles where small pieces of work are *timeboxed*, developed, and released for customer feedback, verification, and validation iteratively. One timebox takes a few weeks up to a month. The agile process model is communication intensive. It eliminates the exponential increase in cost to incorporate changes as in the waterfall model, by keeping the customer involved throughout the development process and validating small pieces of work by them iteratively. These agile process models work better for most of the software projects as changes are inevitable and responding to the change is key to the success of a project.

Since agile development was invented in the mid-1990s, it has revolutionized how software is created by emphasizing short development cycles based on fast customer feedback [32]. As the developers are looking for shorter time period, major new releases are delivered on time. Developers using this methodology call the process *continuous improvement*. But for much of its history, agile development was missing a crucial component: a development platform that supports the rapid development cycles that make the methodology work. In traditional software environments, new software distribution is an ordeal that requires patches, reinstallation, and help from the support team. In such an environment, months or even years are needed to get a new distribution into the hands of users. Incorporating their feedback into the next release then requires comparable time.

6.9.4 Advantages of Agile Model

Agile software process offers the following advantages compared to traditional software development models [33]:

1. *Faster time to market*: Since the software is developed using lesser time in the agile process model, it reduces the time an organization takes to launch the product into the market.
2. *Quick ROI*: Since an organization is able launch the product in lesser time, it generates quick ROI.
3. *Shorter release cycles*: Agile process ensures that the software product is released in shorter cycles compared to traditional software development models.
4. *Better quality*: Since the agile development model ensures the maximum interaction among the stakeholders during the entire process of development, it increases the overall quality of the product.
5. *Better adaptability and responsiveness to business changing requirements*: Since the agile process model is adaptive to incorporate changes in the requirements any time during the development process, it increases the responsiveness to changing requirements of the business.
6. *Early detection of failure/failing projects*: Agile process model involves the maximum interaction among the stakeholders, and the testing phase is not delayed till the entire software development process is complete. This helps in the early detection of failure/failing projects.

6.9.5 How Cloud Meets Agile Process?

The cloud development use case encompasses the flow of defects/requirements through phases of development/builds/tests and back to submission of new requirements or defects by various stakeholders. Automation at any point possible is a key capability, including the ability to *turn on* and *rip down* virtual or physical systems as needed, in a cloud. Continuous integration is a key concept to agile practices. It is based on the philosophy of why wait until the end of the project to see if all pieces of the system will work? Every few hours the system should be fully integrated, but tested with all the latest changes, so the adjustments can be made [34].

It is here that cloud computing makes a substantial difference [32]. Cloud computing eliminates the cumbersome distribution requirements that can bring agile development to a crawl. There are no patches to distribute and no reinstallations needed. With cloud computing, new distributions are installed on hosted servers and made available to users immediately. As a result, it is possible that the application being run today was modified just the night before. One of the best examples of bringing together agile development and cloud computing is the experience of Salesforce.com where, in late 2006, the R&D team moved to agile development.

6.9.5.1 Six Ways the Cloud Enhances Agile Software Development

Cloud computing and virtualization allow the creation of VMs and use of cloud-based services for project management, issue management, and software builds with automated testing. This, in turn, encourages agile development in six key ways. Cloud computing and virtualization make it easy for agile development teams to seamlessly combine multiple development, test, and production environments with other cloud services. Here are six important ways in which cloud computing and virtualization enhance agile software development [35]:

1. *Cloud computing provides an unlimited number of testing and staging servers*: When agile development is used without virtualization or clouds, development teams are limited to one physical server per development, staging, and production server need. However, when VMs or cloud instances are used, development teams have practically an unlimited number of servers available to them. They do not need to wait for physical servers to become free to begin or continue their work.

2. *It turns agile development into a truly parallel activity*: Even in agile development, a developer may experience delays in provisioning server instances and in installing necessary underlying platforms such as database software. Agile development teams can provision the servers they need quickly themselves, rather than wait for IT operations to do it for them.

3. *It encourages innovation and experimentation*: Being able to spawn as many instances as needed enables agile development groups to innovate. If a feature or a story looks interesting, a team can spawn a development instance quickly to code it and test it out. There is no need to wait for the next build or release, as is the case when a limited number of physical servers are available. When adding cloud computing to agile development, builds are faster and less painful, which encourages experimentation.

4. *It enhances continuous integration and delivery*: Having a large number of VMs available to the agile development group in its own cloud or on the public cloud greatly enhances the speed of continuous integration and delivery.

5. *It makes more development platforms and external services available*: Agile development groups may need to use a variety of project management, issue management, and, if continuous integration is used, automated testing environments. A number of these services are available as Software as a Service (SaaS) offerings in the cloud:

   a. Agile development can use a combination of virtualization, private clouds, and the public cloud at the IaaS level. Such offerings include Amazon Web Services, GoGrid, OpSource, and RackSpace Cloud.

   b. Then comes the use of PaaS instances such as the Oracle Database Cloud Service, the Google App Engine, and the Salesforce.com's platform (force.com), all of which include databases and language environments as services.

   c. Finally, there are a number of SaaS services that specifically assist agile development, including Saleforce.com, the Basecamp project management portal, and TestFlight, which provides hosted testing automation for Apple iOS devices.

6. *It eases code branching and merging*: In code refactoring efforts, current releases may need to be enhanced with minor enhancements and used in production, all while a major redesign of code is going on. Code branching is necessary in these cases. Code branching and merging involve juggling many versions of development and staging builds. With virtualization and cloud computing, buying or renting additional physical servers for these purposes can be avoided.

6.9.5.2 Case Study of Agile Development

Meanwhile, Salesforce.com's R&D leverages cloud computing to vastly speed up release cycles [32]. The company's cloud infrastructure helps it maintain a single, unified code base that geographically distributed development teams can use. Those teams are successfully combining agile development and continuous integration/delivery with cloud computing. In reference [32], Salesforce.com finds that agile process model works better on cloud computing platform. Before the introduction of cloud computing, there was a gap or time interval between the releasing of software and getting feedback from the customer and now new software release can be uploaded to the server and used by the customer simultaneously. So, the agile development model can complement the benefits of software services hosted on the Internet. In the rapidly varying computing environment with web services and cloud platform, software design and development also involve various platforms, distributed web services, and geographically distributed enterprises [36].

Salesforce.com's R&D organization has benefitted in several ways from its transition to agile development [32]:

- Increased delivery rate and created a process that makes customers and R&D happy
- Increased time to market of major releases by 61%
- Achieved a Net Promoter Score of 94%, a good indicator of customer satisfaction
- Convinced 90% of the R&D team to recommend the methodology to colleagues inside and outside the company
- Increased productivity across the organization by 38%, as measured by the number of features produced per developer (a side benefit not anticipated as part of the original goals)

6.10 Programming Models

Programming models for cloud computing have become a research focus recently. Cloud computing promises to provide on-demand and flexible IT services, which goes beyond traditional programming models and calls for new ones [37]. Cloud platforms allow programmers to write applications that run in the cloud, or use services from the cloud, or both while abstracting the essence of scalability and distributed processing [38]. With the emergence of cloud as a nascent architecture, abstractions that support emerging programming models are needed. In recent years, cloud computing has led to the design and development of diverse programming models for massive data processing and computation-intensive applications.

Specifically, a programming model is an abstraction of the underlying computer system that allows for the expression of both algorithms and data structures [39]. In comparison, languages and APIs provide implementation of the abstractions and allow algorithms and data structures to be put into practice. A programming model exists independently of the choice of both the programming language and the supporting APIs. Programming models are typically focused on achieving increased developer productivity, performance, and portability to other system designs. The rapidly changing nature of processor architectures and the complexity of designing a platform provide significant challenges for these goals. Several other factors are likely to impact the design of future programming models. In particular, the representation and management of increasing levels of parallelism, concurrency, and memory hierarchies, combined with the ability to maintain a progressive level of interoperability with today's applications, are of significant concern. Furthermore, the successful implementation of a programming model is dependent on exposed features of the runtime software layers and features of the OS [39].

Over the years, many organizations have built large-scale systems to meet the increasing demands of high storage and processing requirements of compute- and data-intensive applications [38]. With the popularity and demands on DCs, it is a challenge to provide a proper programming model that is able to support convenient access to large-scale data for performing computations while hiding all low-level details of physical environments. Cloud programming is about knowing what and how to program on cloud platforms. Cloud platforms provide the basic local functions that an application program requires. These can include an underlying OS and local support such as deployment, management, and monitoring.

6.10.1 Programming Models in Cloud

There are different programming models that are used for solving various compute- or data-intensive problems in cloud. The model to be selected depends on the nature of the problem and also on the QoS expected from the cloud environment. Some of the cloud programming models are discussed in the following subsections.

6.10.1.1 BSP Model

With the advantages on predictable performance, easy programming, and deadlock avoidance, the bulk synchronous parallel (BSP) model has been widely applied in parallel databases, search engines, and scientific computing. The BSP model can be adapted into the cloud environment [37]. The scheduling of computing tasks and the allocation of cloud resources are integrated into the BSP model. Recently, research on cloud computing programming models has made some significant progress, such as Google's MapReduce [40] and Microsoft's Dryad [41]. The BSP model is originally proposed by Harvard's Valiant. Its initial aim is to bridge parallel computation software and architecture. It offers the following advantages: firstly, its performance can be predicted; secondly, no deadlock occurs during message passing; and thirdly, it is easy to program. The BSP model can be used not only for data-intensive applications but also for computation-intensive and I/O-intensive applications.

6.10.1.2 MapReduce Model

Recently, many large-scale computer systems are built in order to meet the high storage and processing demands of compute- and data-intensive applications. MapReduce is one of the most popular programming models designed to support the development of such applications [42]. It was initially created by Google for simplifying the development of large-scale web search applications in DCs and has been proposed to form the basis of a *data center computer*. With the increasing popularity of DCs, it is a challenge to provide a proper programming model that is able to support convenient access to the large-scale data for performing computations while hiding all low-level details of physical environments. Among all the candidates, MapReduce is one of the most popular programming models designed for this purpose.

MapReduce is triggered by the map and reduce operations in functional languages, such as Lisp. This model abstracts computation problems through two functions: map and reduce. All problems formulated in this way can be parallelized automatically. Essentially, the MapReduce model allows users to write map/reduce components with functional-style code. These components are then composed as a dataflow graph to explicitly specify their parallelism. Finally, the MapReduce runtime system schedules these components to distributed resources for execution while handling many tough problems: parallelization, network communication, and fault tolerance.

A map function takes a key/value pair as input and produces a list of key/value pairs as output. A reduce function takes a key and associated value list as input and generates a list of new values as output. A MapReduce application is executed in a parallel manner through two phases. In the first phase, all map operations can be executed independently from each other. In the second phase, each reduce operation may depend on the outputs generated by any number of map operations. All reduce operations can also be executed independently similar to map operations.

The task execution is carried out in four stages: map, sort, merge, and reduce. The map phase is fed with a set of key/value pairs. For each pair, the mapper module generates a result. The sort and merge phases group the data to produce an array, in which each element is a group of values for each key. The reduce phase works on this data and applies the reduce function on it. The hash functions used in the map and reduce functions are user defined and varies with the application of the model. The overall computation is depicted in Figure 6.5.

MapReduce has emerged as an important data-parallel programming model for data-intensive computing [38]. However, most of the implementations of MapReduce are tightly coupled with the infrastructure. There have been programming models proposed that provide a high-level programming interface, thereby providing the ability to create distributed applications in an infrastructure-independent way. Simple API for Grid Applications

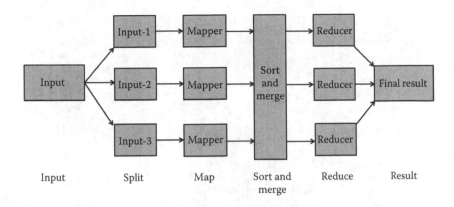

FIGURE 6.5
Computation of MapReduce. (From Jayaraj, A. et al., *Programming Models for Clouds.*)

(SAGA) [43] and Transformer [44] are examples of such models, which try to implement parallel models like MapReduce [42] and All-Pairs [45], taking considerable burden off the application developer.

6.10.1.3 SAGA

Although MapReduce has emerged as an important data-parallel programming model for data-intensive computing, most, if not all, implementations of MapReduce are tightly coupled to a specific infrastructure. SAGA is a high-level programming interface that provides the ability to create distributed applications in an infrastructure-independent way [38]. SAGA supports different programming models and concentrates on the interoperability on grid and cloud infrastructures. SAGA supports job submission across different distributed platforms, file access/transfer, and logical file, as well as checkpoint recovery and service discovery. SAGA API is written in C++ and supports other languages like Python, C, and Java. The runtime environment decision making is given support by the engine that loads relevant adaptors, as shown in Figure 6.6.

6.10.1.4 Transformer

Even though there are existing programming models based on C++ and Java in the industrial market, they suffer from certain shortcomings. First, the programmers have to master the bulky and complex APIs in order to use the model. Secondly, most programming models are designed for specific programming abstractions and created to address one particular kind of problem. There is an absence of a universal distributed software framework for processing massive datasets. To address the aforementioned shortcomings, a new framework called Transformer [38] is used, which supports diverse programming models and also is not problem specific.

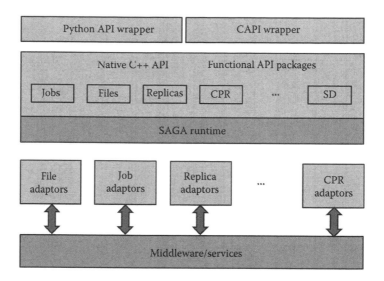

FIGURE 6.6
SAGA model. (From Jayaraj, A. et al., *Programming Models for Clouds*.)

Transformer is based on two concise operations: send and receive. Using the Transformer model, various models such as MapReduce [39], Dryad, and All-Pairs [40] can be built. The architecture of the Transformer model is divided into two layers: common runtime and model-specific systems, as shown in Figure 6.7. This is done to reduce coupling. Runtime system handles the tasks-like flow of data between machines and executes the tasks on different systems making use of send and receive functions from runtime API. Model-specific layer deals with particular model tasks like mapping, data partitioning, and data dependencies.

Transformer has a master/slave architecture. Every node has two communication components: a message sender and a message receiver. The

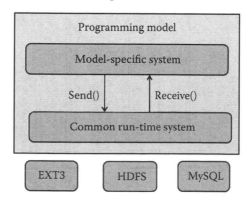

FIGURE 6.7
Transformer architecture. (From Jayaraj, A. et al., *Programming Models for Clouds*.)

master node issues commands for task execution on the slave nodes. The slave nodes return the status of execution when it is over. The fault-tolerance strategy is agile in nature. Failure is detected by the runtime system whereas fault recovery is handled by the model-specific layer. This involves rerunning the tasks or resending data. Transformer system is coded in Python. Communication between nodes is done using message-passing mechanism opposed to semaphores and conditions in threaded approach. Since the frequency of communication is high, asynchronous network programming is adopted, and moreover the message is serialized before sending it. Using the Transformer model, all three known parallel programming models, namely MapReduce, Dryad, and All-Pairs, are implemented.

6.10.1.5 Grid Batch Framework

Recently, an alternative to parallel computational models has been suggested that enables users to partition their data in a simplified manner while having the highest possible efficiency. The Grid Batch system has two fundamental data types [38]: table and indexed table. A table is a set of rows that are independent of each other. An indexed table has all the properties of a table in addition to having an index associated with each record.

The two major software components of the Grid Batch system are the Distributed File System (DFS) and the Job Scheduler. The DFS is responsible for storing and managing the files across all the nodes in the system. A file is broken down into many pieces and each of these pieces is stored on a separate node. The Job Scheduler constitutes of a master node and associated slave nodes. A job is broken down into many smaller tasks by the master node, and each of these tasks is distributed among the slave nodes. The basic map and reduce operators in the MapReduce system are extended in the Grid Batch model. These are map operator, distribute operator, join operator, Cartesian operator, recurse operator, and neighbor operator.

6.11 Pervasive Computing

Pervasive computing is a combination of technologies, such as Internet capabilities, voice recognition, networking, artificial intelligence, and wireless computing, used to make computing anywhere possible. Pervasive computing devices make day-to-day computing activities extremely easy to perform. The technology is moving beyond the PC to everyday devices with embedded technology and connectivity. Pervasive computing is also called ubiquitous computing, in which almost any device or material such as clothing, tools, appliances, vehicles, homes, human body, or even the coffee mug can be imbedded with chips to connect that object to an infinite network of other devices. The goal of pervasive computing, which combines current network technologies with wireless computing, voice recognition, Internet capability, and artificial intelligence, is to create an environment where the connectivity of devices is achieved in such a way that the connectivity is unobtrusive and always available. Pervasive computing also has a number of prospective applications, which range from home care and health, to geographical tracking and intelligent transport systems.

The words pervasive and ubiquitous mean *existing everywhere*. Pervasive computing devices are completely connected and constantly available. Pervasive computing relies on the convergence of wireless technologies, advanced electronics, and the Internet. The goal of researchers working in pervasive computing is to create smart products that communicate unobtrusively. The products are connected to the Internet and the data they generate are easily available. An example of a practical application of pervasive computing is the replacement of old electric meters with smart meters. In the past, electric meters had to be manually read by a company representative. Smart meters report usage of electricity in real time over the Internet. They will also notify the power company when there is an outage and also send messages to display units in the home and regulate the water heater.

Hence, in pervasive computing, computing is made to appear everywhere and anywhere [46]. In contrast to desktop computing, pervasive computing can be done using any supporting device, at any location. The underlying technologies to support pervasive computing include Internet, advanced middleware, OS, mobile, sensors, microprocessors, new I/O and user interfaces, networks, mobile protocols, and location-based services. This paradigm is also named with different names like physical computing, the Internet of Things, and things that think, by considering the objects involved in it. In this case, the device used to access applications and information is almost irrelevant as various types of devices or platform can be used to perform the intended operation [47].

6.11.1 How Pervasive Computing Works?

The success of ubiquitous computing rests with the proper integration of various components that talk to each other and thereby behaving as a single connected system. Figure 6.8 shows the architecture of a ubiquitous computing stack [48]. At the bottom of the stack is a *physical* layer. Tiny sensors are attached (carried, worn, or embedded) to people, animals, machines, homes, cars, buildings, campuses, and fields. Sensors capture various bits of information from the immediate surroundings. Beyond the microphone and camera, multiple sensors such as GPS, accelerometer, and compass can be integrated into it.

Above the sensors lies the wireless communication infrastructure, which can be provided by the 802.11 family of networks. Together with mesh networks, such standards ensure the connectivity of sensors and devices.

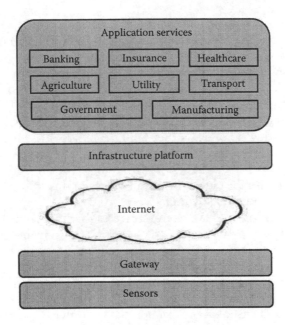

FIGURE 6.8
Pervasive computing stack. (From Perspectives, *TCS Consult. J.*, available at: http://sites.tcs. com/insights/perspectives/enterprise-mobility-ubiquitous-computing-beyond-mobility#. UzPo8fmSzCe.)

Another technology called ZigBee is a low-cost alternative for keeping multiple devices connected, allowing parent devices to wirelessly control child sensors. Near field communication (NFC) is yet another technology standard that leverages RFID and can be used for ubiquitous computing, especially in scenarios where non-battery-operated passive points are concerned. NFC-powered devices can also interact with one another.

The next level includes a range of application services. The data from the sensors and handheld devices are gathered, mined, and analyzed for patterns. The patterns help provide options to smart applications that proactively make changes to environments through smartphones, tablets, notebooks, or any other handheld devices or smart devices. An example could be that of a cardiac patient wearing a tiny monitor connected to a mobile device. An irregular ECG will trigger the mobile to alert the patient's doctor and emergency services.

6.11.2 How Pervasive Computing Helps Cloud Computing?

Nowadays, IT enterprises are adopting cloud computing in order to reduce the total cost involved and also to improve the QoS delivered to the customers. Cloud computing provides the opportunity to access the infrastructure, platform, and the software from the service providers on a pay-per-use basis. Pervasive computing helps cloud computing by providing the ability to access the cloud resources anytime, anywhere and also through any device. Pervasive computing provides the necessary features such as ubiquitous computing, storage and archiving, social community–based applications, and business as well as nonbusiness applications in order for cloud computing to gain its full potential. Cloud computing is typically a client-server architecture, where the client can be any portable device like a laptop, phone, browser, or any other OS-enabled devices [49]. A main issue with these portable devices is the constraints they present in terms of storage, memory, processing, and battery lifetime. By storing data on the cloud, and interacting with the cloud through secure communication channels, all these constraints can be easily met.

Machine-to-machine (M2M) communication is important in cloud computing [47]. Such a communication scenario spans from the shop floor, to the DC, to the boardroom, as the devices carried along track the user's movements and activities and also interact with the other systems around. For example, an employee is currently in New York and he wants to discuss something with two colleagues. He requests an appointment using his mobile device, and based on his location data and that of his colleagues, and the timing of the meeting, backend systems automatically book him a conference room and set up a video link to a coworker out of town. Based on analytics and the title of the meeting, relevant documents are dropped into a collaboration space. The employee's device records the meeting to an archive and notes who has attended in person. And, this conversation is automatically transcribed, tagged, and forwarded to team members for review.

Wearable devices like Google Glass will also feed into the new workplace. The true power behind these applications is not in the devices themselves but in the analytic systems that back them. The backend systems of the applications combine the data collected from the various types of computing devices such as Google Glass, smartphones, an embedded GPS device in a palette, or a sensor in a car's engine. Such systems process the data and then turn it into useful information that is used for triggering the required actions. Different computing systems performing various activities are deployed with APIs so that the user can build applications that extract information from these multiple systems.

6.12 Operating System

An OS is a collection of softwares that manages the computer hardware resources and other programs in the computing system. It provides common services required by computer programs for their effective execution within the computing environment. The OS is an essential component of the system software in a computer system as application programs usually require an OS for their interface with the hardware resources and other system programs. For hardware functions such as input and output, and memory allocation, the OS acts as an intermediary between programs and the computer hardware.

6.12.1 Types of Operating Systems

The different variants of OSs are the following:

1. *Network OSs*: A network operating system (NOS) is a computer OS that is designed primarily to support workstations, PCs that are connected on a LAN. An NOS provides features such as printer sharing, common file system and database sharing, application sharing, security mechanisms, and also the ability to manage a network name directory and other housekeeping functions of the network. Novell's NetWare and Microsoft's LAN Manager are examples of NOSs. In addition, some multipurpose OSs, such as Windows NT and Digital's OpenVMS, come with capabilities that enable them to be described as an NOS.

2. *Web OSs*: Web OSs are basically websites that replicate the desktop environment of modern OSs, all inside a web browser. They are installed onto web servers and live on the Internet. Thus, a user can access his virtual desktop from any device, anywhere, that is connected to the net. Web OSs are also called the dynamic computers. In this case, the applications, hard disk, and OSs are all present at the servers from where they are accessed. The web OS service provider manages the application and database accesses of the various users. The user is provided with a graphical user interface similar to the one available on a desktop PC, which can be used to access the data and the applications from the server. Google Chrome OS is an example of a web OS.

3. *Distributed OS*: A distributed OS is a software that is present over a collection of independent, networked, communicating, and physically separate computational nodes. Each individual node holds a specific software that is a subset of the global aggregate OS. Each subset consists of two distinct components of the distributed OS. The first one is a ubiquitous minimal kernel, or microkernel, that directly controls the node's hardware. The second one is a higher-level collection of system management components that coordinate the node's individual and collaborative activities. The microkernel and the management components work together. They support the distributed system's goal of integrating multiple resources and processing functionality into an efficient and stable system. To a user,

a distributed OS works in a manner similar to a single-node, mono-lithic OS. That is, although it consists of multiple nodes, it appears to the users and applications as a single-node OS.

4. *Embedded systems*: Embedded systems are OSs present in elec-tronic devices used for various purposes in order to make them *smart* and more efficient. Embedded systems present in devices such as routers, for example, typically include a preconfigured web server, DHCP server, and some utilities for its effective networking operation, and they do not allow the installation of new programs in them. Examples of embedded OSs for routers include Cisco Internetwork Operating System (IOS), DD-WRT, and Juniper Junos. An embedded OS can also be found inside an increasing number of consumer gadgets including phones, personal digital assistance (PDA), and digital media player for the successful completion of their intended tasks.

6.12.2 Role of OS in Cloud Computing

In the 1970s, International Business Machines Corporation (IBM) released an OS called VM that allowed mainframe systems to have multiple vir-tual systems, or VMs on a single physical node [50]. The VM OS materi-alized shared access of mainframe systems to the next level by allowing multiple distinct computing environments to live in the same physical environment. Most of the basic functions of current virtualization soft-ware are inherited from this early VM OS. The virtualization software is now represented with the term *hypervisor*. Hardware virtualization is used to share the resources of the cloud providers effectively with the customers by generating an illusion of dedicated computing, storage, and networking on a computing infrastructure. The concept of virtualization is physically implemented using the hypervisor modules, and the opera-tion and processing of hypervisors are materialized by the OSs. In other words, hypervisor modules are installed on top of OSs, which act as an interface between hardware units and hypervisor packages.

Using virtualization, multiple VMs are generated according to the require-ment, and these VMs are made operational by individually installing OSs on each VM. Figure 6.9 shows the virtualization of a single hardware of the CSP to create different VMs, each installed with its own OS. Every VM runs custom OS or guest OS that has its own memory, CPU, and hard drives along with CD-ROMs, keyboards, and networking, despite the fact that all of those resources are shared among the VMs.

In addition, an OS such as Linux supports the necessary standards that enhance portability and interoperability across cloud environments [51]. OS platforms are designed to hide much of the complexity required to

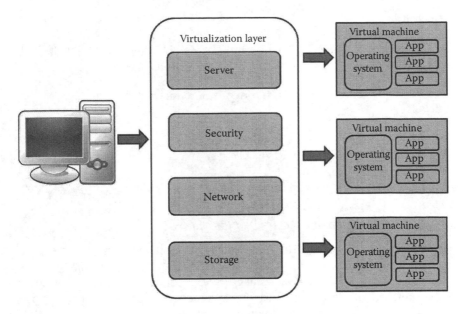

FIGURE 6.9
OS and virtualization. (From Steddum, J., A brief history of cloud computing, available at: http://blog.softlayer.com/tag/mainframe.)

support applications running in complex and federated environments. It needs to work effectively in the background in order to ensure that all the right resources (such as processing power, required memory, and storage) are allocated when needed. In addition, the OS implements the level of security and QoS to ensure that applications are able to access the resources needed to deliver an acceptable level of performance, in an efficient and secure manner.

One of the most important ways to support the underlying complexity of well-managed cloud computing resources is through the OS [51]. One of the most significant requirements for companies adopting cloud computing is the need to adopt a hybrid approach to computing. To do so, most organizations will continue to maintain their traditional DC to support complex mixed workloads. For example, an organization may choose a public cloud environment for development and test workloads, a private cloud for customer-facing web environments that deal with personal information, and a traditional DC for legacy billing and financial workloads. It is considered that hybrid cloud computing environments will be the norm for the future. Therefore, it is more important than ever for the OS to support and federate the various computing deployment models so that they appear to be a single system from a customer experience and a systems management perspective.

6.12.3 Features of Cloud OS

The elements required to create an operationally sophisticated hybrid cloud computing environment include the following [51]:

1. Well-defined interfaces that hide implementation details
2. Core security services
3. The ability to manage virtualization
4. Management of workloads to provide QoS and performance

These features are explained in the following subsections.

6.12.3.1 Well-Defined and Abstracted Interfaces

A cloud OS should provide the APIs that enable data and services interoperability across distributed cloud environments. Mature OSs provide a rich set of services to the applications so that each application does not have to invent important functions such as VM monitoring, scheduling, security, power management, and memory management. In addition, if APIs are built on open standards, it will help organizations avoid vendor lock-in and thereby creating a more flexible environment. For example, linkages will be required to bridge traditional DCs and public or private cloud environments. The flexibility of movement of data or information across these systems demands the OS to provide a secure and consistent foundation to reap the real advantages offered by the cloud computing environments. Also, the OS needs to make sure the right resources are allocated to the requesting applications. This requirement is even more important in hybrid cloud environments. Therefore, any well-designed cloud environment must have well-defined APIs that allow an application or a service to be plugged into the cloud easily. These interfaces need to be based on open standards to protect customers from being locked into one vendor's cloud environment.

6.12.3.2 Support for Security at the Core

Whether an organization is considering a public, private, or hybrid cloud environment, security is the most important foundation to ensure the protection for its assets. Security issues are exacerbated in a cloud environment since it is highly distributed, and also it involves a large variety of internal and external systems added or removed to/from the cloud dynamically. Therefore, the cloud environment has to protect the identity of the users and their information from external threats. To support the needs of the organizations, cloud security requires an integrated management capability within the OS that can track all IT assets in the context of how they are being used. This capability needs to ensure that the security meets an organization's compliance and governance requirements.

Both virtualization and multitenancy supports have to be implemented in a secure manner. As virtualization and multitenancy become the norm in cloud environments, it is critical that security be built-in at the core. When servers are virtualized, it makes the creation of a new image possible with little effort. This expansion of virtual images raises the risk of attack because it increases the possibility that a security flaw in the hypervisor can be exploited by a guest instance. It can expose both existing systems and other partners that interact with those systems to security threats. When security is implemented as a framework within the OS, it improves the overall security of both virtualized and nonvirtualized environments.

6.12.3.3 Managing Virtualized Workloads

Virtualization is fundamental to cloud computing because it breaks the traditional links between the physical server and the application. These virtualized environments are controlled and managed by a hypervisor. In essence, the hypervisor is an OS on the physical hardware and presents the core hardware abstraction and I/O instructions needed by the guests in its environment. Hence, the effective management of the hypervisor and the virtualized environments is critical to the success of cloud computing.

6.12.3.4 Management of Workloads

The cloud environment has to be designed in a manner that protects the individual customer's workloads. Hence, the prerequisite for effective cloud computing is the ability to isolate workloads of users from each other. The cloud OS should make sure that all the required resources are effectively managed in the cloud environment for the individual users.

6.12.4 Cloud OS Requirements

Other than the features of the cloud OS explained in the previous section, the major OS requirements in a cloud environment are given in the following [52]:

1. *The cloud OS must permit autonomous management of its resources*: The cloud OS should expose a consistent and unified interface that conceals whenever possible the fact that individual nodes are involved in its operations, and what those low-level operations are. It should support the autonomous management of the various cloud resources on behalf of its users and applications.

2. *Cloud OS operation must continue despite failure of nodes, clusters, and network partitioning*: Guaranteeing continued operation of the cloud management processes in these conditions involves mechanisms for

quickly detecting the failures and enacting appropriate measures for recovering from the failures. Several cloud libraries that implement common fault tolerance and state recovery features are provided for the customers to use.

3. *The cloud must support multiple types of applications*: Applications of different types such as high-performance computing, high data availability, and high network throughput should ideally coexist in a cloud environment and obtain from the system the resources that best match the application requirements.

4. *The cloud OS management system must be decentralized, scalable, and cost effective*: Moreover, apart from initial resource deployment, no human intervention should be required to expand the cloud resources. Likewise, user management should only entail the on-demand creation of user credentials, which are then automatically propagated throughout the cloud.

5. *The resources used in the cloud architecture must be accountable*: The resource usage of the various cloud customers should be monitored effectively in the cloud environment. This monitoring activity could be used for charging the customers for their resource access, and also for the security auditing (if needed). Moreover, dynamic billing schemes based on resource congestion could be an effective way for resource allocation.

6.12.5 Cloud-Based OS

Researchers are now aiming to go one step further and take the OS to the cloud with TransOS, a cross-platform, cloud-based OS [53]. The TransOS system code is stored on a cloud server and a minimal amount of code would be required to boot up the computer and connect it to the Internet. Featuring a graphical user interface, TransOS downloads specific pieces of code to perform the same kinds of tasks as a conventional OS, thereby allowing a bare bones terminal to perform tasks beyond the limitations of its hardware. The terminal would make a call to the relevant TransOS code as and when required, ensuring that the inactive OS is not hogging system resources when applications are being run. The TransOS manages all the networked and virtualized hardware and software resources and enables the users to select and run any service on demand. The TransOS could be adapted to platforms other than PCs such as mobile devices, factory equipment, and even domestic appliances.

In addition to keeping a lean machine, TransOS users can also store their documents and files in the cloud, much like Apple's iCloud, keeping their local storage free. With TransOS, users never have to worry about running the most up-to-date version of the OS or even maintain their own computer. With their OS, data, files, and settings stored in the cloud, any computer with an Internet connection (e.g., computers that are publicly available at libraries and colleges) becomes just like the user's own machine. The files stored within the cloud can also be accessed anytime from any Internet-ready device, including smartphones and tablets.

6.13 Application Environment

An application development environment (ADE) is the hardware, software, and computing resources required for building software applications. ADE is a composite set of computing resources that provides an interface for application development, testing, deployment, integration, troubleshooting, and maintenance services. These are combined with the software engineering resources, such as a programming language's integrated development environment (IDE), reporting and analysis software, troubleshooting tools, and other performance evaluation software utilities.

6.13.1 Need for Effective ADE

As the mobile web application market matures, competitive pressures and user expectations will drive application developers to differentiate their product offerings by providing value-added features [54]. The standards-based application environment must ensure the interoperability of the application development components. In particular, it must enable customized content, extensible functionality and advanced user interfaces.

For the web to be ubiquitous, web access devices must be present almost everywhere. For this to occur, web access devices must become small and portable. As web-enabled devices evolve from today's desktop computers to such things as cellular telephones, car radios, and personal organizers, the challenge will be to provide a common application authoring environment across a diverse range of devices. The existing *standard* web application environment consists of HTML, JavaScript, and an ad hoc collection of standard graphics file formats, processed by an HTML browser. To incorporate multimedia content and extend the expressiveness or the functionality of a user interface, Java applets and browser plug-ins can be used. They require extensions that are often device specific and require special installation.

Hence, the web application environment must provide application developers the tools they need to develop innovative products, without sacrificing interoperability. Hence, an effective ADE should

- Support multiple content types, including multimedia content
- Support multimodal user interaction
- Provide a framework for the integration of new technologies as they become available

The application environment should support a standard extensibility framework. As new media types, user agents, or supplemental services emerge, they should be integrated into the environment in a backward-compatible manner without affecting the performance of any existing applications.

6.13.2 Application Development Methodologies

Today, two development methodologies are widely used in application development: distributed and agile developments [55].

6.13.2.1 Distributed Development

This is the natural by-product of the Internet and the phenomenon that not all coding geniuses live within commuting distance from the workplace. Distributed development is a global development that brings its own challenges with collaboration and code management. There are applications available for distributed code management such as git and Subversion. They are widely used in distributed environments.

6.13.2.2 Agile Development

This is where cloud development can really be much more than just *online*. Since cloud environments can be provisioned instantly and nearly any configuration can be copied and activated, the possibilities for instant developments and test environments are very attractive to developers. Cloud development can also boost agile development by coordinating collaboration, planned sprints, and emergency bug fixes. Deploying to the cloud is also very useful for agile development. Prereleases can be pushed out to customers' test machines on their cloud almost instantly. Even if the customer is not in a cloud environment yet, prereleases can be posted on a public cloud for the customer to access and test remotely before accepting delivery of the final release of the application. Toolsets that can help the agile management in the cloud include Code2Cloud, in conjunction with Tasktop and CollabNet.

6.13.3 Power of Cloud Computing in Application Development

Cloud computing has effectively solved the financial and infrastructural problems associated with developing custom applications for the enterprises as it eases the financial investment that was previously required to set up the sophisticated developer environment necessary to build, test, and deploy custom applications in-house [56]. As a result, the introduction of cloud platforms has enabled developers to solely focus on creating highly scalable modern applications. Further, the process of marketing these custom applications is less time consuming and more effective as a result of the flexibility provided by cloud computing services. When applications are run in the cloud, they are accessed as a service—this is known as Software as a Service (SaaS). By utilizing SaaS, the companies can deliver services in a cost-effective and efficient manner. This process enables businesses to work in conjunction with partners to develop applications and quickly distribute them in the market.

The advantages of using cloud computing services over traditional software go beyond just the drop in costs. The traditional methods of developing custom applications that often took months to complete has now dropped to just weeks. With all the required software and tools available in the cloud, developers can work more efficiently and productively than they could if they were using traditional software, where, more often than not, additional components were required to develop a complete application. Today's heavily simplified approach of accessing applications online allows developers to produce comprehensive enterprise-level applications simply through a web browser, without the technical difficulties associated with traditional solutions.

Another main benefit of using cloud computing services for application development is the efficient use of resources. Applications that utilize virtualized IT services are generally more efficient and better equipped to meet user demands. The pay-per-use model of cloud computing services provides the clients with flexibility to spend according to their requirements and thus eliminates the unnecessary expenses. Also, cloud computing services allow delivering the applications on multiple devices; this allows companies to design their applications so that they are compatible with a range of devices.

6.13.3.1 Disadvantages of Desktop Development

Desktop development environments are becoming outdated, failing more often, and causing productivity issues for developers. The main issues with desktop environment are the following [57]:

1. *Complicated configuration management*: The substantial configuration management process for a developer's workspace turns developers into part-time system administrators, responsible for their own mini-DC running entirely on the desktop. This is time consuming, error prone, and challenging to automate. Many developers have multiple computers and are forced to repeat these tasks on each machine. There is no way to synchronize the configurations of components across different machines, and each machine requires similar hardware and OSs to operate the components identically.

2. *Decreased productivity*: Many IDEs are memory and disk hogs, with significant boot times. They are so resource-hungry that they can starve other applications and the net effect is less productivity due to a slower machine.

3. *Limited accessibility*: Normally, desktop developer workspaces are not accessible via mobile devices through the Internet. Developers who need remote access have to resort to some complex and slow solutions such as *GotoMyPC*.

4. *Poor collaboration*: These days, most developers work as part of a team, so communication and collaboration among the team members are critical for the success of the project. In the case of desktop IDEs, they must outsource collaboration to communication systems outside the developer's workflow, forcing developers to continuously switch between developing within the IDE and communicating with their team via other means. To solve these problems, it requires moving the entire development workspace into the cloud. The cloud-based environment is centralized, making it easy to share. Developers can invite others into their workspace to coedit, cobuild, or codebug and can communicate with one another in the workspace itself. The cloud can offer improvements in system efficiency, giving each individual workspace a configurable slice of the available memory and computing resources.

6.13.3.2 Advantages of Application Development in the Cloud

Cloud platforms reduce the overall development time of a software project [58]. This is largely due to the cloud platform's ability to streamline the development process, including the ability to quickly get the development assets online. Moreover, cloud platforms provide the ability to collaborate effectively on development efforts. Cloud-based development platforms in PaaS and IaaS public clouds such as Google, Amazon Web Services, Microsoft, and Salesforce.com offer cost savings and better QoS.

Some of the benefits of the application development in the cloud are given as follows:

- The ability to self-provision development and testing environments
- The ability to quickly get applications into production and to scale those applications as required
- The ability to collaborate with other developers, architects, and designers on the development of the application

6.13.4 Cloud Application Development Platforms

Application development, deployment, and runtime management have always been reliant on development platforms such as Microsoft's .NET, WebSphere, or JBoss, which have been deployed on premise traditionally [59]. In the cloud computing context, applications are generally deployed by cloud providers to provide highly scalable and elastic services to as many end users as possible. Cloud computing infrastructure needs to support many users to access and utilize the same application services, with elastic allocation of resources. This has led to the enhancement in development platform technologies and architectures to handle performance, security, resource allocation, application monitoring, billing, and fault tolerance. Cloud provides the ADE as PaaS. There are several solutions available in the PaaS market, including Google App Engine, Microsoft Windows Azure, Force.com, and Manjrasoft Aneka.

6.13.4.1 Windows Azure

Windows Azure provides a wide array of Windows-based services for developing and deploying Windows-based applications on the cloud. It makes use of the infrastructure provided by Microsoft to host these services and scale them seamlessly. The Windows Azure Platform consists of SQL Azure and the .NET services. The .NET services comprise of access control services and .NET service bus. Windows Azure is a platform with shared multitenant hardware provided by Microsoft. Windows Azure application development mandates the use of SQL Azure for RDBMS functionality, because that is the only coexisting DBMS functionality accessible in the same hardware context as the applications.

6.13.4.2 Google App Engine

Google App Engine provides an extensible runtime environment for web-based applications developed with Java or Python, which leverage huge Google IT infrastructure. Google App Engine is offered by Google, Inc. Its key value is that developers can rapidly build web-based applications on their machine and deploy them on the cloud. Google App Engine provides developers with a simulated environment to build and test applications locally with any OS or any system that runs a suitable version of Python and Java language environments. Google uses the JVM with Jetty Servlet engine and Java Data Objects.

6.13.4.3 Force.com

Force.com is a development and execution environment and is the best approach for PaaS for developing customer relationship management (CRM)–based applications. With regard to the design of its platform and the runtime environment, it is based on the Java technology. The platform uses a proprietary programming language and environment called Apex code, which has a reputation for simplicity in learning and rapid development and execution.

6.13.4.4 Manjrasoft Aneka

Aneka is a distributed application platform for developing cloud applications. Aneka can seam together any number of Windows-based physical or virtual desktops or servers into a network of interconnected nodes that act as a single logical *application execution layer*. Aneka-based clouds can be deployed on a variety of hardware and OSs including several flavors of the Windows and Linux OS families. Aneka provides a flexible model for developing distributed applications and provides integration with external clouds such as Amazon EC2 and GoGrid. Aneka offers the possibility to select the most appropriate infrastructure deployment without being tied to any specific vendor, thus allowing enterprises to comfortably scale to the cloud as and when needed.

6.13.5 Cloud Computing APIs

APIs are provided by some of the CSPs for the development of cloud applications. Details of some of the APIs provided by the CSPs such as Rackspace, IBM, and Intel are given in the following [60].

6.13.5.1 Rackspace

Developers have access to the API documentation and software development kit (SDK) across all of Rackspace's services at their developer site, http://developer.rackspace.com. Thus, Rackspace provides developers with the tools and resources necessary to create new applications and services on top of their APIs.

6.13.5.2 IBM

IBM introduced new APIs, which can be found at the IBM developer site, www.ibm.com/developerworks/. The introduction of the new APIs focuses on arming developers with the tools and resources to build new products, applications, and services.

6.13.5.3 Intel

Intel has several SDKs that aimed at cloud computing developers. Intel has a cloud services platform beta where developers can download the SDK for identity-based and cross-platform services. The Intel Cloud Builders program brings together leading systems and software solutions vendors to provide the best practices and practical guidance on how to deploy, maintain, and optimize a cloud infrastructure based on Intel architecture. And for developers seeking to use public cloud infrastructure services, the Intel Cloud Finder makes it easier to select providers that meet a developer's requirements.

6.14 Summary

Cloud computing is dominating the IT industry worldwide today. More and more companies and organizations are adopting the cloud model these days. Even though cloud computing is a new service delivery model, the underlying technologies have been in existence for a long time. Cloud computing uses many of those technologies to achieve its established goals. This chapter focuses on the various technological drivers of cloud computing. It discusses about the basic enabling technologies of cloud computing such as SOA, hypervisors and virtualization, multicore technology, and memory and storage technologies. It also talks about the latest developments in Web 2.0 and Web 3.0, the advancements in the programming models, software development models, pervasive computing, OSs, and ADEs. It also explains how these technologies are related to the cloud model, helping the cloud in delivering quality services. The recent developments in each of these enabling technologies are highlighted with their advantages and characteristic features. The chapter explains as to how these underlying technologies are empowering the present cloud computing paradigm to deliver its services effectively. Also, the chapter presents how various stakeholders such as service providers and service consumers are benefitted from the features extended by these technologies.

Review Points

- *SOA*: Service-oriented architecture is a flexible set of design principles and standards used for systems development and integration. A properly implemented SOA-based system provides a loosely coupled set of services that can be used by the service consumers for meeting their service requirements within various business domains (see Section 6.2).

- *Hypervisor*: Hypervisors are software tools used to create virtual machines, and they produce the virtualization of various hardware resources such as CPU, storage, and networking devices. They are also called virtual machine monitor (VMM) or virtualization managers (see Section 6.3.2).

- *Multicore technology*: In the multicore technology, two or more CPUs are working together on the same chip. In this type of architecture, a single physical processor contains the core logic of two or more processors (see Section 6.4).

- *Storage as a Service*: Storage as a Service (STaaS) is a cloud business model in which a service provider rents space in its storage infrastructure to various cloud users (see Section 6.5.3).

- *Software-defined networking*: Software-defined networking (SDN) is an approach to networking in which control is decoupled from networking hardware and given to a software application called the controller (see Section 6.6.5).

- *Web 2.0*: Web 2.0 (or Web 2) is the popular term given to the advanced Internet technology and applications that include blogs, wikis, RSS, and social bookmarking (see Section 6.7).

- *Semantic web*: The semantic web is a vision of IT that allows data and information to be readily interpreted by machines, so that the machines are able to take contextual decisions on their own by finding, combining, and acting upon relevant information on the web (see Section 6.8.1.1).

- *Agile development model*: Agile model is a software development model where the software is developed in rapid, incremental cycles. The development results in tiny incremental releases and is based on previously built functionality and is carefully tested to ensure software quality (see Section 6.9.3).

- *MapReduce*: MapReduce is a popular programming model designed to support the development of compute- and data-intensive applications, which requires high storage and processing demands (see Section 6.10.1.2).

- *Pervasive computing*: Pervasive computing is a combination of technologies such as Internet capabilities, voice recognition, networking, artificial intelligence, and wireless computing used to make computing anywhere possible (see Section 6.11).

- *Web OS*: Web operating systems are basically websites that replicate the desktop environment of modern OSs, all inside a web browser (see Section 6.12.1).

- *Cloud API*: Cloud APIs are provided by the cloud service providers for the development of cloud applications (see Section 6.13.5).

Review Questions

1. What are the characteristic features of SOA that are used in the successful deployment of cloud computing?

2. What are the various approaches in virtualization? What are the roles played by the hypervisor and virtualization in cloud environment?

3. How can the multicore technologies be used to achieve the parallelism in cloud?

4. What are the latest technological developments to meet the storage requirements in cloud?

5. How does SDN relate to the cloud computing scenario?

6. What are the ways in which cloud computing relies on the concepts of Web 2.0 for its successful operation?

7. How do semantic web and web services contribute to the evolution of cloud computing?

8. Justify the decision to adopt the agile development model for software development. How can the cloud computing paradigm make the agile process effective?

9. What are the programming models used in cloud? Justify the answer by explaining the characteristic features of the models.

10. Explain the ways in which pervasive computing affects the cloud model.

11. Explain the differences between a web OS and a cloud OS.

12. How does the cloud computing paradigm help in effective application development?

References

1. Strassmann, P. A. How SOA fits into cloud computing. *SOA Symposium*, April 22, 2010.
2. Zhang, L.-J. and Q. Zhou. CCOA: Cloud computing open architecture. *IEEE International Conference on Web Services 2009 (ICWS 2009)*, 2009, pp. 607–616.
3. Gschwind, M. *Multicore Computing and the Cloud: Optimizing Systems with Virtualization*. IBM Corporation, 2009.
4. Sun, X.-H., Y. Chen, and S. Byna. Scalable computing in the multicore era. *Proceedings of the International Symposium on Parallel Architectures, Algorithms and Programming*, 2008.
5. Sankaralingam, K. and R. H. Arpaci-Dusseau. Get the parallelism out of my loud. *Proceedings of the Second USENIX Conference on Hot Topics in Parallelism*, 2010.
6. Jamal, M. H. et al. Virtual machine scalability on multi-core processors based servers for cloud computing workloads. *IEEE International Conference on Networking, Architecture, and Storage, 2009 (NAS 2009)*, Hunan, China, July 9–11, 2009. IEEE, New York, 2009, pp. 90–97.
7. Venkatraman, A. Intel launches micro-server, network, storage technologies to power cloud datacenters. Available at: http://www.computerweekly.com/news/2240204767/Intel-launches-micro-server-network-storage-technologies-to-power-cloud-datacentres. Accessed December 10, 2013.
8. Introduction to storage technologies. Consulting Solutions, White Paper, Citrix XenDesktop.

9. Cloud storage for cloud computing. OpenGrid Forum, SNIA, Advancing Storage and Information Technology, White Paper, 2009.
10. Stryer, P. Understanding data centers and cloud computing. Global Knowledge Instructor, CCSI, CCNA.
11. Ingthorsson, O. Networking technologies in cloud computing. Available at: http://cloudcomputingtopics.com/2010/04/networking-technologies-in-cloud-computing/#comments. Accessed December 13, 2013.
12. Bitar, N., S. Gringeri, and T. J. Xia. Technologies and protocols for data center and cloud networking. *IEEE Communications Magazine* 51(9): 24–31, 2013.
13. Rosen, E. and Y. Rekhter. BGP/MPLS IP virtual private networks (VPNs). RFC 4364, 2006.
14. Sajassi et al. BGP MPLS based ethernet VPN. Work in progress, 2013.
15. O'Reilly, T. What is Web 2.0. O'Reilly Network, 2005. Accessed August 6, 2006.
16. VMWARE. Virtualization overview. Available at: www.vmware.com.
17. Techtarget. Definition of Web 2.0. Available at: http://whatis.techtarget.com/definition/Web-20-or-Web-2. Accessed December 1, 2013.
18. Reservoir Consortium. Resources and services virtualization without barriers. Scientific report. 2009.
19. Mulholland, A., J. Pyke, and P. Finger. *Enterprise Cloud Computing: A Strategy Guide for Business and Technology*, Meghan-Kiffer Press, Tampa, FL.
20. Keen, A., Web 1.0 + Web 2.0 = Web 3.0, Typepad.com. Available at: http://andrewkeen.typepad.com/the-great-seduction/2008/04/web-10-web20-w.html. Accessed November 21, 2013.
21. Viluda, P., Differences between Web 3.0 and Web 2.0 standards. Available at: http://www.cruzine.com/2011/02/14/web-3-web-2-standards/.
22. World Wide Web Consortium (W3C). W3C semantic web activity, 2011. Retrieved November 26, 2011. Accessed November 26, 2011.
23. Getting, B. Basic definitions: Web 1.0, Web. 2.0, Web 3.0. Available at: http://www.practicalecommerce.com/articles/464-Basic-Definitions-Web-1-0-Web-2-0-Web-3-0. Accessed December 1, 2013.
24. Spivack, N. Web 3.0: The third generation web is coming. Available at: http://lifeboat.com/ex/web.3.0. Accessed December 1, 2013.
25. Hoy, T. Web 3.0: Converging cloud computing and the web. Available at: http://www.ebizq.net/topics/cloud_computing/features/12477.html?page=3. Accessed November 23, 2013.
26. Shaw, T. Web 3.0 gives business smarter infrastructure. Available at: http://www.baselinemag.com/cloud-computing/Web-30--Gives-Business-Smarter-Infrastructure/. Accessed November 27, 2013.
27. Sommerville, I. *Software Engineering*, 8th edn. Pearson Education, 2006.
28. Guha, R. and D. Al-Dabass. Impact of Web 2.0 and cloud computing platform on software engineering. *2010 International Symposium on Electronic System Design (ISED)*, Bhubaneswar, India, December 20–22, 2010. IEEE, New York, 2010, pp. 213–218.
29. Pressman, R. *Software Engineering: A Practitioner's Approach*, 7th edn. McGraw-Hill Higher Education, New York, 2009.
30. Singh, A., M. Korupolu, and D. Mahapatra. Server-storage virtualization: Integration and load balancing in data centers. *International Conference for High Performance Computing, Networking, Storage and Analysis, 2008 (SC 2008)*, Austin, TX, November 15–21, 2008. IEEE/ACM Supercomputing (SC), 2008, pp. 1–12.

31. Velagapudi, M. SDLC for cloud computing—How is it different from the traditional SDLC? Available at: http://blog.bootstraptoday.com/2012/02/06/sdlc-for-cloud-computing-how-is-it-different-from-the-traditional-sdlc/.
32. Salesforce.com. Agile development meets Cloud computing for extraordinary results. Available at: www.salesforce.com. Accessed October 3, 2013.
33. Dumbre, A., S. S. Ghag, and S. P. Senthil. Practising Agile software development on the Windows Azure platform. Infosys Whitepaper, 2011.
34. Gulrajani, N. and D. Bowler. *Software Development in the Cloud–Cloud Management and ALM.*
35. Kannan, N. Ways the cloud enhances agile software development. Available at: http://www.cio.com/article/714210/6_Ways_the_Cloud_Enhances_Agile_Software_Development. Accessed December 5, 2013.
36. Mahmood, Z. and S. Saeed. *Software Engineering Frameworks for Cloud Computing Paradigm.* Springer-Verlag, London, U.K., 2013.
37. Liu, X. A programming model for the cloud platform. *International Journal of Advanced Science and Technology* 57: 75–81, 2013.
38. Jayaraj, A., J. John Geevarghese, K. Rajan, U. Kartha, and V. Samuel Varghese. *Programming Models for Clouds.*
39. McCormick, P. et al. Programming models. White Paper. Available at: https://asc.llnl.gov/exascale/exascale-pmWG.pdf.
40. Dean, J. and S. Ghemawat. Mapreduce: Simplified data processing on large clusters. *Communications of the ACM* 51(1): 107–113, 2008. Accessed December 8, 2013.
41. Iiard, M., M. Budiu, and Y. Yuan. Dryad: Distributed data-parallel programs from sequential building blocks. *Operating Systems Review* 41(3): 59–72, 2007.
42. Jin, C. and R. Buyya. Mapreduce programming model for. NET-based cloud computing. In: H. Sips, D. Epema, and H.-X. Lin (eds.), *Euro-Par 2009 Parallel Processing.* Springer, Berlin, Germany, 2009, pp. 417–428.
43. Miceli, C. et al. Programming abstractions for data intensive computing on clouds and grids. *9th IEEE/ACM International Symposium on Cluster Computing and the Grid,* 2009.
44. Wang, P. et al. *Transformer: A New Paradigm for Building Data-Parallel Programming Models.* IEEE Computer Society, 2009.
45. Gannon, D. The computational data center—A science cloud. Indiana University, Bloomington, IN.
46. Soylu, A., P. De Causmaecker, and P. Desmet. Context and adaptivity in pervasive computing environments: Links with software engineering and ontological engineering. *Journal of Software* 4(9): 992–1013, 2009.
47. Gallagher, S. Forget "post-PC"—Pervasive computing and cloud will change the nature of IT. Available at: http://arstechnica.com/information-technology/2013/08/forget-post-pc-pervasive-computing-and-cloud-will-change-the-nature-of-it/2/. Accessed October 24, 2013.
48. Perspectives. Ubiquitous computing: Beyond mobility: Everywhere and every thing. *TCS Consulting Journal.* Available at: http://sites.tcs.com/insights/perspectives/enterprise-mobility-ubiquitous-computing-beyond-mobility#.UzPo8fmSzCe.
49. Namboodiri, V. Sustainable pervasive computing through mobile clouds. Available at: http://sensorlab.cs.dartmouth.edu/NSFPervasiveComputingAtScale/pdf/1569391485.pdf. Accessed October 28, 2013.

50. Steddum, J. A brief history of cloud computing. Available at: http://blog.softlayer.com/tag/mainframe.
51. Hurwitz, J. The role of the operating system in cloud environments. A Hurwitz White Paper, 2011.
52. Pianese, F. et al. Toward a cloud operating system. *Network Operations and Management Symposium Workshops (NOMS Wksps), 2010 IEEE/IFIP.* IEEE, 2010.
53. Quick, D. Cloud-based operating system in the works. Available at: http://www.gizmag.com/transos-cloud-based-operarting-system/24494/. Accessed December 14, 2013.
54. Dominiak, D. Standardizing a web-based application environment. Motorola White Paper. Available at: http://www.w3.org/2000/09/Papers/Motorola.html. Accessed December 1, 2013.
55. Proffitt, B. Building applications in the cloud: A tour of the tools. Available at: http://www.itworld.com/virtualization/189811/building-applications-cloud-tour-tools. Accessed November 24, 2013.
56. Siddiqui, Z. The impact of cloud computing on custom application development. Available at: http://www.trackvia.com/blog/technology/cloud_computing_and_custom_application_developments. Accessed December 6, 2013.
57. Linthicum, D. Why application development is better in the cloud. Available at: http://www.infoworld.com/d/cloud-computing/why-application-development-better-in-the-cloud-211239. Accessed December 10, 2013.
58. Jewell, T. Why cloud development environments are better than desktop development. Available at: http://readwrite.com/2013/04/16/why-cloud-development-environments-are-better-than-desktop-development#awesm=~ozHQROKsJUAOI2. Accessed December 4, 2013.
59. Buyya, R. and K. Sukumar. Platforms for building and deploying applications for cloud computing. arXiv preprint arXiv:1104.4379, 2011.
60. Le, T. Developers and cloud computing application programming interfaces (APIs). Available at: http://software.intel.com/en-us/blogs/2013/09/26/developers-and-cloud-computing-application-programming-interfaces-apis.

Further Reading

Dean, J. and S. Ghemawat. MapReduce: Simplified data processing on large clusters. *Proceedings of the Sixth Symposium on Operating System Design and Implementation,* 2004.

7

Virtualization

Learning Objectives

The main objective of this chapter is to introduce the concept of virtualization and how it is used as enabling technology for cloud computing. After reading this chapter, you will

- Understand the basics of virtualization
- Understand how the different resources such as processors, memory, storage, and network can be virtualized
- Understand the pros and cons of different approaches to virtualization
- Understand the basics of hypervisor and its security issues
- Understand how cloud computing is different from virtualization
- Understand how cloud computing leverages the virtualization for its different service models

Preamble

Virtualization is an enabling technology for the different cloud computing services. It helps to improve scalability and resource utilization of the underlying infrastructure. It also enables the IT personnel to perform the administration task easier. With the help of resource sharing, the hypervisor supports the green IT services. This chapter describes virtualization and discusses the benefits of virtualization and, different resources that can be virtualized. This chapter also explains the different approach for virtualization such as full virtualization, hardware-assisted virtualization, and paravirtualization. The different types of hypervisors and its security issues are also discussed. At the end of the chapter, the difference between cloud computing and virtualization and how virtualization is used by cloud computing to provide services are discussed.

7.1 Introduction

In recent years, computing becomes more complex and requires large infrastructure. The organizations invest huge amount on buying additional physical infrastructure as and when there is a need for more computing resources. If you look at the capital expenditure (CapEx) and operational expenditure (OpEx) of buying and maintaining large infrastructure, it is really high. At the same time, the resource utilization and return on investment (ROI) on buying the additional infrastructure are very low. To increase the resource utilization and ROI, the companies started using the technology called *virtualization* where a single physical infrastructure can be used to run multiple operating systems (OSs) and applications. The virtualization is not a new word to the computing world; it is being used for at least past four decades. The term *virtualization* becomes a buzzword in recent years and most organizations started to adopt it rapidly because of its benefits like efficient resource utilization and increased ROI, ease of administration, and green IT support.

Virtualization is a technology that enables the single physical infrastructure to function as a multiple logical infrastructure or resources. Virtualization is not only limited to the hardware, it can take many forms such as memory, processor, I/O, network, OS, data, and application. The different forms of virtualization will be discussed in the next section.

Before virtualization, the single physical infrastructure was used to run a single OS and its applications, which results in underutilization of resources. The nonshared nature of the hardware forces the organizations to buy a new hardware to meet their additional computing needs. For example, if any organization wants to experiment or simulate their new idea, they have to use separate dedicated systems for different experiments. So to complete their research work successfully, they tend to buy a new hardware that will increase the CapEx and OpEx. Sometimes, if the organization does not have money to invest more on the additional resources, they may not be able to carry out some valuable experiments because of lack of resources. So, people started thinking about sharing a single infrastructure for multiple purposes in the form of virtualization. The computing scenarios before and after virtualization are shown in Figures 7.1 and 7.2, respectively.

After virtualization was introduced, different OSs and applications were able to share a single physical infrastructure. The virtualization reduces the huge amount invested in buying additional resources. The virtualization becomes a key driver in the IT industry, especially in cloud computing. Generally, the terms *cloud computing* and *virtualization* are not same. There are significant differences between these two technologies, which will be

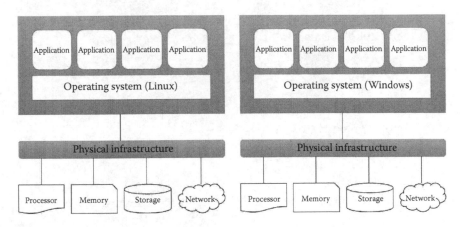

FIGURE 7.1
Before virtualization.

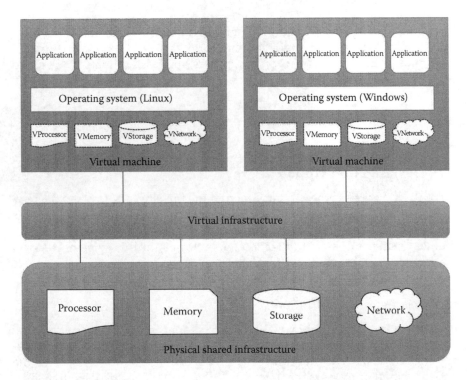

FIGURE 7.2
After virtualization.

discussed in the later part of this chapter. Industries adopt virtualization in their organization because of the following benefits:

- Better resource utilization
- Increases ROI
- Dynamic data center
- Supports green IT
- Eases administration
- Improves disaster recovery

While virtualization offers many benefits, it also has some drawbacks:

- Single point of failure
- Demands high-end and powerful infrastructure
- May lead to lower performance
- Requires specialized skill set

This chapter focuses on the different virtualization opportunities, different approaches to virtualization, role of the hypervisors in virtualization, attacks that target the hypervisors, and virtualization for cloud computing.

7.2 Virtualization Opportunities

Virtualization is the process of abstracting the physical resources to the pool of virtual resources that can be given to any virtual machines (VMs). The different resources like memory, processors, storage, and network can be virtualized using proper virtualization technologies. In this section, we shall discuss some of the resources that can be virtualized.

7.2.1 Processor Virtualization

Processor virtualization allows the VMs to share the virtual processors that are abstracted from the physical processors available at the underlying infrastructure. The virtualization layer abstracts the physical processor to the pool of virtual processors that is shared by the VMs. The virtualization layer will be normally any hypervisors. Processor virtualization from a single hardware is illustrated in Figure 7.3. But processor virtualization can also be achieved from distributed servers.

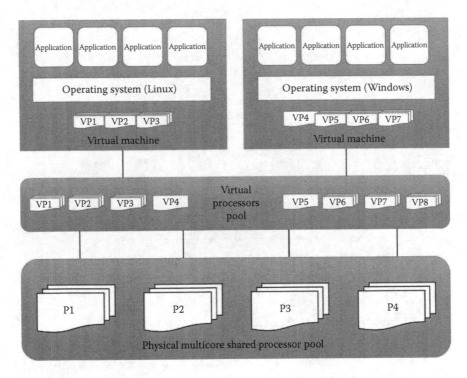

FIGURE 7.3
Processor virtualization.

7.2.2 Memory Virtualization

Another important resource virtualization technique is memory virtualization. The process of providing a virtual main memory to the VMs is known as memory virtualization or main memory virtualization. In main memory virtualization, the physical main memory is mapped to the virtual main memory as in the virtual memory concepts in most of the OSs. The main idea of main memory virtualization is to map the virtual page numbers to the physical page numbers. All the modern x86 processors are supporting main memory virtualization.

Main memory virtualization can also be achieved by using the hypervisor software. Normally, in the virtualized data centers, the unused main memory of the different servers will consolidate as a virtual main memory pool and can be given to the VMs. The concept of main memory virtualization is illustrated in Figure 7.4.

7.2.3 Storage Virtualization

Storage virtualization is a form of resource virtualization where multiple physical storage disks are abstracted as a pool of virtual storage disks to

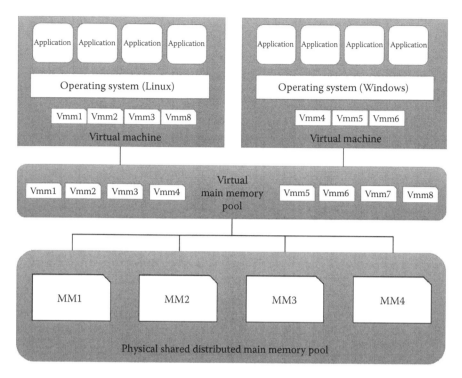

FIGURE 7.4
Main memory virtualization.

the VMs. Normally, the virtualized storage will be called a logical storage. Figure 7.5 illustrates the process of storage virtualization.

Storage virtualization is mainly used for maintaining a backup or replica of the data that are stored on the VMs. It can be further extended to support the high availability of the data. It can also be achieved through the hypervisors. It efficiently utilizes the underlying physical storage. The other advanced storage virtualization techniques are storage area networks (SAN) and network-attached storage (NAS).

7.2.4 Network Virtualization

Network virtualization is a type of resource virtualization in which the physical network can be abstracted to create a virtual network. Normally, the physical network components like router, switch, and Network Interface Card (NIC) will be controlled by the virtualization software to provide virtual network components. The virtual network is a single software-based entity that contains the network hardware and software resources. Network virtualization can be achieved from internal network or by combining many external networks. The other advantage

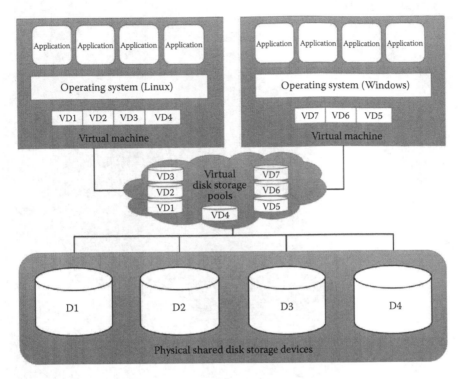

FIGURE 7.5
Storage virtualization.

of network virtualization is it enables the communication between the VMs that share the physical network. There are different types of network access given to the VMs such as bridged network, network address translation (NAT), and host only. The concept of network virtualization is illustrated in Figure 7.6.

7.2.5 Data Virtualization

Data virtualization is the ability to retrieve the data without knowing its type and the physical location where it is stored. It aggregates the heterogeneous data from the different sources to a single logical/virtual volume of data. This logical data can be accessed from any applications such as web services, E-commerce applications, web portals, Software as a Service (SaaS) applications, and mobile application.

Data virtualization hides the type of the data and the location of the data for the application that access it. It also ensures the single point access to data by aggregating data from different sources. It is mainly used in data integration, business intelligence, and cloud computing. Figure 7.7 represents data virtualization technology.

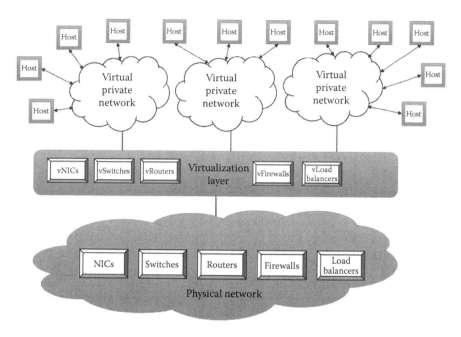

FIGURE 7.6
Network virtualization.

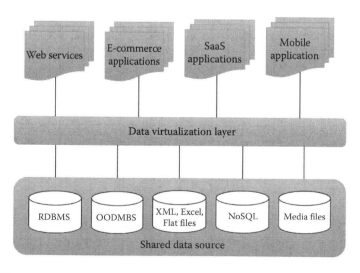

FIGURE 7.7
Data virtualization.

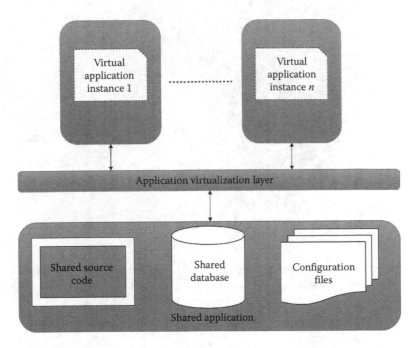

FIGURE 7.8
Application virtualization.

7.2.6 Application Virtualization

Application virtualization is the enabling technology for SaaS of cloud computing. The application virtualization offers the ability to the user to use the application without the need to install any software or tools in the machine. Here, the complexity of installing the client tools or other supported software is reduced. Normally, the applications will be developed and hosted in the central server. The hosted application will be again virtualized, and the users will be given the separated/isolated virtual copy to access. The concept of application virtualization is illustrated in Figure 7.8.

7.3 Approaches to Virtualization

There are three different approaches to virtualization. Before discussing them, it is important to know about *protection rings* in OSs. Protection rings are used to isolate the OS from untrusted user applications. The OS can be protected with different privilege levels. In protection ring architecture, the rings are arranged in hierarchical order from ring 0 to ring 3 as shown in Figure 7.9. Ring 0 contains the programs that are most privileged, and ring 3 contains the

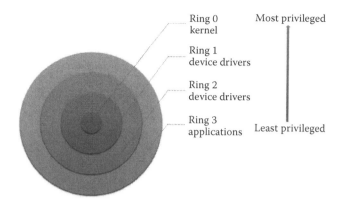

FIGURE 7.9
Protection rings in OSs.

programs that are least privileged. Normally, the highly trusted OS instructions will run in ring 0, and it has unrestricted access to physical resources. Ring 3 contains the untrusted user applications, and it has restricted access to physical resources. The other two rings (ring 1 and ring 2) are allotted for device drivers. This protection ring architecture restricts the misuse of resources and malicious behavior of untrusted user-level programs. For example, any user application from ring 3 cannot directly access any physical resources as it is the least privileged level. But the kernel of the OS at ring 0 can directly access the physical resources as it is the most privileged level.

Depending on the type of virtualization, the hypervisor and guest OS will run in different privilege levels. Normally, the hypervisor will run with the most privileged level at ring 0, and the guest OS will run at the least privileged level than the hypervisor. There are three types of approaches followed for virtualization:

1. Full virtualization
2. Paravirtualization
3. Hardware-assisted virtualization

Each of the virtualization approaches is discussed in detail in this section.

7.3.1 Full Virtualization

In full virtualization, the guest OS is completely abstracted from the underlying infrastructure. The virtualization layer or virtual machine manager (VMM) fully decouples the guest OS from the underlying infrastructure. The guest OS is not aware that it is virtualized and thinks it is running on the real hardware. In this approach, the hypervisor or VMM resides at ring 0 and provides all the virtual infrastructures needed for VMs.

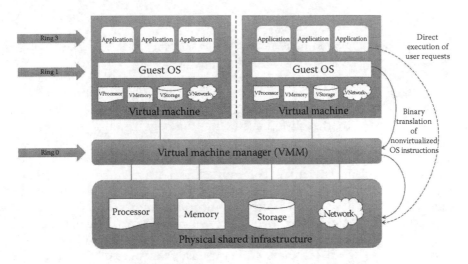

FIGURE 7.10
Full virtualization.

The guest OS resides at ring 1 and has the least privilege than the hypervisor. Hence, the OS cannot communicate to the physical infrastructure directly. It requires the help of hypervisors to communicate with the underlying infrastructure. The user applications reside at ring 3, as shown in Figure 7.10. This approach uses binary translation and direct execution techniques. Binary translation is used to translate nonvirtualized guest OS instructions with new sequences of instructions that have the same intended effect on the virtual infrastructure. On the other hand, direct execution is used for user application requests where the applications can directly access the physical resources without modifying the instructions.

Pros

- This approach provides the best isolation and security for the VMs.
- Different OSs can run simultaneously.
- The virtual guest OS can be easily migrated to work in native hardware.
- It is easy to install and use and does not require any change in the guest OS.

Cons

- Binary translation is an additional, overhead, and it reduces the overall system performance.
- There is a need for correct combination of hardware and software.

7.3.2 Paravirtualization

This approach is also known as *partial virtualization* or *OS-assisted virtualization* and provides partial simulation of the underlying infrastructure. The main difference between the full virtualization and paravirtualization is the guest OS knows that it is running in virtualized environment in paravirtualization. But in full virtualization, this information is not known to the guest OS. Another difference is that the paravirtualization replaces the translation of nonvirtualized OS requests with *hypercalls*. Hypercalls are similar to system calls and used for the direct communication between OS and hypervisor. This direct communication between the guest OS and hypervisor improves performance and efficiency. In full virtualization, the guest OS will be used without any modification. But in paravirtualization, the guest OS needs to be modified to replace nonvirtualizable instructions with the hypercalls.

As shown in Figure 7.11, the modified guest OS resides at ring 0 and the user applications at ring 3. As the guest OS is at privileged position, it can communicate directly to the virtualization layer without any translation by means of hypercalls. Like in full virtualization, the user applications are allowed to access the underlying infrastructure directly.

Pros

- It eliminates the additional overhead of binary translation and hence improves the overall system efficiency and performance.
- It is easier to implement than full virtualization as there is no need for special hardware.

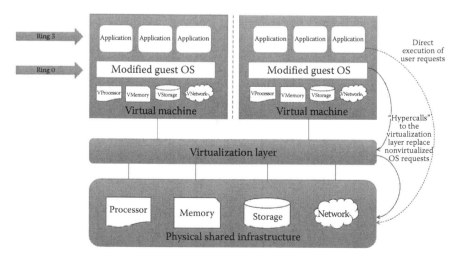

FIGURE 7.11
Paravirtualization.

Cons

- There is an overhead of guest OS kernel modification.
- The modified guest OS cannot be migrated to run on physical hardware.
- VMs suffer from lack of backward compatibility and are difficult to migrate to other hosts.

7.3.3 Hardware-Assisted Virtualization

In the two previous approaches, there is an additional overhead of binary translation or modification of guest OS to achieve virtualization. But in this approach, hardware vendors itself, like Intel and AMD, offer the support for virtualization, which eliminates much overhead involved in the binary translation and guest OS modification. Popular hardware vendors like Intel and AMD has given the hardware extension to their x86-based processor to support virtualization.

For example, the Intel releases its Intel Virtualization Technology (VT-x) and AMD releases its AMD-v to simplify the virtualization techniques. In VT-x, the guest state is stored in *virtual machine control structures* and in AMD-v in *virtual machine control blocks*. In hardware-assisted virtualization, the VMM has the highest privilege (root privilege) level even though it is working below ring 0. The OS resides at ring 0 and the user application at ring 3, as shown in Figure 7.12. Unlike the other virtualization approaches, the guest OS and the user applications are having the same privilege level (nonroot privilege level). As discussed earlier, the hardware-assisted virtualization technique removes binary translation and paravirtualization. Here, the OS requests directly trap the hypervisor without any translation. As in other virtualization approaches, the user requests are directly executed without any translation.

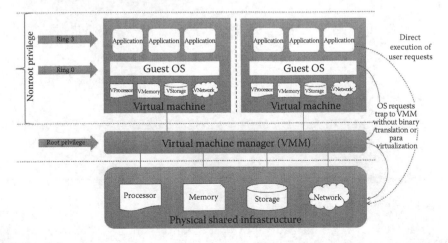

FIGURE 7.12
Hardware-assisted virtualization.

TABLE 7.1

Summary of the Different Approaches to Virtualization

	Full Virtualization	Paravirtualization	Hardware-Assisted Virtualization
Technique	Binary translation and direct execution	Hypercalls	OS requests trap to VMM without binary translation or paravirtualization
Guest OS modification	No	Yes	No
Compatibility	Excellent compatibility	Poor compatibility	Excellent compatibility
Is guest OS hypervisor independent?	Yes	No	Yes
Performance	Good	Better in certain cases	Fair
Position of VMM and privilege level	Ring 0 Root privilege	Below ring 0	Below ring 0 Root privilege
Position of guest OS and privilege level	Ring 1 Nonroot privilege	Ring 0 Root privilege	Ring 0 Nonroot privilege
Popular vendor(s)	VMware ESX	Xen	Microsoft, Virtual Iron, and XenSorce

Pros

- It reduces the additional overhead of binary translation in full virtualization.
- It eliminates the guest OS modification in paravirtualization.

Cons

- Only new-generation processors have these capabilities. All x86/x86_64 processors do not support hardware-assisted virtualization features.
- More number of VM traps result in high CPU overhead, limited scalability, and less efficiency in server consolidation.

A summary of the different approaches to virtualization is given in Table 7.1.

7.4 Hypervisors

VMs are widely used instead of physical machines in the IT industry today. VMs support green IT solutions, and its usage increases resource utilization, making the management tasks easier. Since the VMs are mostly used, the technology that enables the virtual environment also gets attention in industries and academia. The virtual environment can be created with the help of a software tool called *hypervisors*. Hypervisors are the software tool that sits in between VMs and physical infrastructure and provides the required virtual infrastructure for VMs. Generally, the virtual infrastructure means virtual CPUs (vCPUs), virtual memory, virtual NICs (vNICs), virtual storage, and virtual I/O devices. The hypervisors are also called VMM. They are the key drivers in enabling virtualization in cloud data centers. There are different hypervisors that are being used in the IT industry. Some of the examples are VMware, Xen, Hyper-V, KVM, and OpenVZ. The different types of hypervisors, some popular hypervisors in the market, and security issues with recommendations are discussed in this section.

7.4.1 Types of Hypervisors

Before hypervisors are introduced, there was a one-to-one relationship between hardware and OSs. This type of computing results in underutilized resources. After the hypervisors are introduced, it became a one-to-many relationship. With the help of hypervisors, many OSs can run and share a single hardware. Hypervisors are generally classified into two categories:

1. Type 1 or bare metal hypervisors
2. Type 2 or hosted hypervisors

The major difference between these two types of hypervisors is that type 1 runs directly on the hardware and type 2 on host OS. Figures 7.13 and 7.14 illustrate the working of type 1 and type 2 hypervisors, respectively.

Type 1 hypervisor is also known as bare metal or native hypervisor. It can run and access physical resources directly without the help of any host OS. Here, the additional overhead of communicating with the host OS is reduced and offers better efficiency when compared to type 2 hypervisors. This type of hypervisors is used for servers that handle heavy load and require more security. Some examples of type 1 hypervisors include Microsoft Hyper-V, Citrix XenServer, VMWare ESXi, and Oracle VM Server for SPARC.

Type 2 hypervisors are also known as embedded or hosted hypervisors. This type of hypervisors requires the host OS and does not have direct access to the physical hardware. These types of hypervisors are installed on the host OS as a software program. The host OS is also known as physical host, which has the direct access to the underlying hardware. The major disadvantage of this approach is if the host OS fails or crashes, it also results in crashing of VMs. So, it is recommended to use type 2 hypervisors only on client systems where efficiency is less critical. Examples of type 2 hypervisors include VMWare Workstation and Oracle Virtualbox.

A summary of some of the popular hypervisors that are in the market is given in Table 7.2.

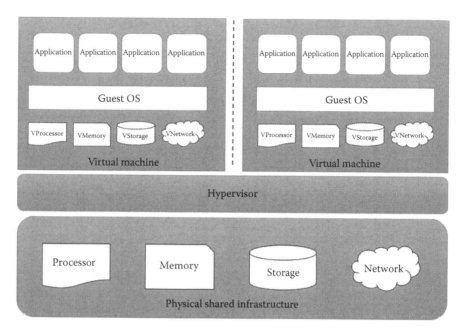

FIGURE 7.13
Type 1 or bare metal hypervisor.

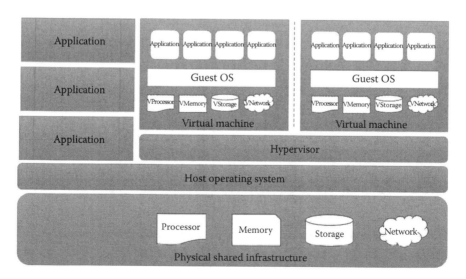

FIGURE 7.14
Type 2 or hosted hypervisor.

TABLE 7.2

Summary of Hypervisors

Hypervisor	Vendor	Type	License
Xen	University of Cambridge Computer Laboratory	Type 1	GNU GPL v2
VMWare ESXi	VMware, Inc.	Type 1	Proprietary
Hyper-V	Microsoft	Type 1	Proprietary
KVM	Open virtualization alliance	Type 2	GNU general public license
VMWare workstation	VMware, Inc.	Type 2	Shareware
Oracle Virtualbox	Oracle Corporation	Type 2	GNU general public license version 2

7.4.2 Security Issues and Recommendations

The hypervisor creates a virtual environment in the data centers. So, the better way to attack the resources is attacking the hypervisor. The hypervisor attack generally compromises the hypervisor through malicious code written by any attacker to disrupt or corrupt the whole server. In a virtualized environment, hypervisor is the higher authority entity that has the direct access to the hardware. So, most of the attackers will target the hypervisor as an entry point to attack the system. In bare metal hypervisor (type 1), it is very difficult to perform the attack as it is deployed directly on the hardware. But the hosted hypervisors (type 2) are more vulnerable to the attacks as hypervisors are running on top of the host OSs. There are two possibilities of attacking the hypervisor:

1. Through the host OS
2. Through the guest OS

Attack through the host OS: Attacks from the host OS can be performed by exploiting the vulnerabilities of the host OS. It is known that even the modern OSs are also vulnerable to the attacks. Once the OS gets compromised, the attackers have full control over the applications running on top of the OS. As hypervisors (type 2) are also an application that is running on top of the OS, there is a possibility of attacking the hypervisor through the compromised host OS. The idea behind attacking the hypervisor through the host OS is illustrated in Figure 7.15. Once the attacker gets full control over the hypervisor through the compromised OS, the attacker will be able to run all the privileged instructions that can control the actual hardware. The attacker can do the following malicious activities:

- Denial of service attack, where the attacker can deny the virtual resources when there is a request from the new VM
- Stealing the confidential information that is stored in the VMs

Attack through the guest OS: The hypervisor can also be compromised or attacked from the malicious script from the compromised guest OS.

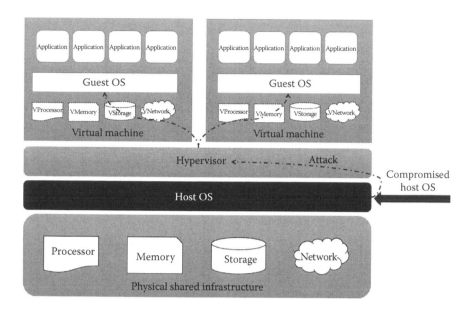

FIGURE 7.15
Attack through the host OS.

Since the guest OS is communicating with the hypervisor to get virtual resources, any malicious code from the guest OS or VMs can compromise the hypervisor. Normally, the attacks from the guest OS will try to abuse the underlying resources. The idea behind the attack through the guest OS is illustrated in Figure 7.16. As shown in the figure, the attacker will try to

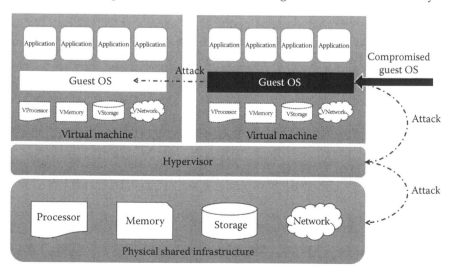

FIGURE 7.16
Attack through the guest OS.

attack or compromise the hypervisor from the malicious VMs. Once the hypervisor gets compromised from the guest OS or malicious VMs, it can misuse the hypervisors' high privilege on the hardware. This type of attack is possible in both type 1 and type 2 hypervisors. After the hypervisor gets compromised, the attacker can do the following malicious activities:

- Get the unauthorized access to the other VMs that share the physical hardware.
- Attacker can utilize the hardware resources fully to launch resource exhaustion attacks, etc.

Recommendations to avoid hypervisor attacks: Most of the attacks on the hypervisor are through the host OS or the guest OS. So, the OS should be protected from the malicious attacker or the hypervisors should be protected from the compromised OSs. There are several best practices to keep the hypervisor secured:

- Update the hypervisor software and the host OS regularly.
- Disconnect the unused physical resources from the host system or hypervisor.
- Enable least privilege to the hypervisor and guest OS to avoid the attacks through unauthorized access.
- Deploy the monitoring tools in the hypervisor to detect/prevent malicious activities.
- Strong guest isolation.
- Employ mandatory access control policies.

7.5 From Virtualization to Cloud Computing

Many users of current IT solutions consider the technologies *virtualization* and *cloud computing* as the same. But both technologies are actually different, or in other words, we can say virtualization is not cloud computing. We can prove this claim with the following parameters:

1. *Type of service*: Generally, virtualization offers more infrastructure services rather than platform and application services. But cloud computing offers all infrastructure (IaaS), platform (PaaS), and software (SaaS) services.
2. *Service delivery*: The service delivery in cloud computing is on-demand and allows the end users to use the cloud services as per the need. But virtualization is not made for on-demand services.

3. *Service provisioning*: In cloud computing, automated and self-service provisioning is possible for the end users, whereas in virtualization, it is not possible and a lot of manual work is required from the providers or system administrator to provide services to the end users.

4. *Service orchestration*: Cloud computing allows the service orchestration and service composition to meet end user requirements. Some providers are also providing automated service orchestration to the end users. But in virtualization, orchestrating different service to get composite services is not possible.

5. *Elasticity*: One of the important characteristics that differentiate cloud computing from virtualization is elasticity. In cloud computing, we can add or remove the infrastructure dynamically according to the need, and adding or removing the infrastructure is automatic. But virtualization fails to provide elasticity as stopping and starting a VM is manual and is also difficult.

6. *Targeted audience*: The targeted audience of these two technologies is also different. Cloud computing targets the service providers for high resource utilization and improved ROI. At the same time, it also facilitates the end users to save money by using on-demand services. In the case of virtualization, the targeted audience is only the service providers or IT owners, not the end users.

With this short discussion, we can conclude that *cloud computing and virtualization are different*. But there might be some question that will arise from the IT owners: "I already invested more in virtualization technology, do I need to change everything to get the benefits of cloud computing?" The answer to this question is *no* as cloud computing uses virtualization for its service delivery. Cloud computing can run on any virtualized environment as virtualization is one of the enabling technologies for cloud computing. Of course, without virtualization, cloud computing might not exist. Cloud computing uses virtualization for better resource utilization and is coupled with utility computing to benefit service providers, developers, and end users. In other words, we can say that cloud computing takes virtualization to the next step. Cloud computing and virtualization converge in better resource utilization, and virtualization stops there whereas cloud computing moves one step ahead and joins utility computing to provide IT as a service. There are different cloud service models available, namely Infrastructure as a Service (IaaS), Platform as a Service (PaaS), and SaaS. In this section, we shall discuss how cloud computing uses the virtualization technology to provide different cloud services.

7.5.1 IaaS

The cloud computing service delivery model that allows the customers to access the resources as a service from the service provider data center is known as the Infrastructure as a Service (IaaS) model. The virtualization concept is fully utilized in the infrastructure layer of the cloud computing. The IaaS service offers virtual memory, virtual processors, virtual storage, and virtual networks to run the VMs.

The IaaS service utilizes the memory, processor, storage, and network virtualization of the underlying infrastructure. The IaaS layer uses the hypervisors to abstract the underlying resources for the VMs. The virtual data center will not be simply referred to as a cloud data center. The virtualized data center will be called as cloud data center if it delivers the service on a pay-per-use basis. Normally, for achieving IaaS services, type 1 hypervisors will be selected rather than type 2 hypervisors as the type 1 hypervisors are directly accessing the underlying hardware. IaaS is generally provided in the form of VMs that uses the virtual resources abstracted from the physical resources.

Normally, the real cloud data centers will contain the networked server machines for providing massive infrastructure services. So, whenever one server is overloaded with many VMs, the additional request will be migrated to the other free physical server by using the load balancer. There are many IaaS providers that are available in the market, which include Amazon, Microsoft, OpenStack, Eucalyptus, and CloudStack. The general service provisioning mechanism of IaaS services is illustrated in Figure 7.17.

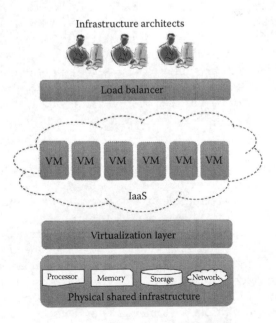

FIGURE 7.17
IaaS.

7.5.2 PaaS

The Platform as a Service (PaaS) allows the end user to develop and deploy the application online by using the virtual development platform provided by the service provider. Generally, the service provider will provide all the development tools as a service to the end users through the Internet. The end users need not install any integrated development environments (IDEs), programming languages, and component libraries in their machine to access the services.

The programming languages, databases, language runtimes, middleware, and component libraries will be provided to the customers by abstracting the actual platform that runs in the provider data center. Generally, the deployment of application developed using the PaaS service depends on the type of cloud deployment model. If the end users select any public cloud deployment model, the application will be hosted as an off-premise application. If the users select the private deployment model, the application will be accessed as an on-premise application. Generally, the PaaS services utilize the OS-level, database-level, programming language–level virtualization to provide the virtual development platform to the end users. Generally, the PaaS providers will provide a variety of client tools such as WebCLI, REST APIs, and Web UI to the developers for accessing the virtual platform. Some PaaS providers allow the offline development by integrating with the IDEs like eclipse to make the development environment availability. The developers need not be online to use their services. They can work offline and push the application online whenever it is ready for the deployment. Here, application scalability is a very important factor. The scalability of the application can be achieved by the proper load balancer that transfers the extra load to the new server. Examples of PaaS service providers include Google App Engine, Microsoft Windows Azure, Redhat OpenShift, and force.com. A general overview of the PaaS service is illustrated in Figure 7.18.

7.5.3 SaaS

Like infrastructure and platform, software applications can also be virtualized. The software delivery model that allows the customers to access the software that is hosted in the service provider data center through the Internet is known as Software as a Service (SaaS).

Generally, SaaS is a subscription-based application rather than a licensed application. To access the SaaS application, customers need not install it on their machine. With the simple web browser, they can access the application from the service provider data center through the Internet. SaaS utilizes application-level virtualization to deploy the application. The SaaS application allows multiple customers to share the same instance of an application. This technology is popularly known as multitenancy. Since many users are sharing the application, the load on the application will

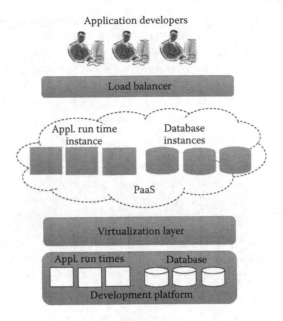

FIGURE 7.18
PaaS.

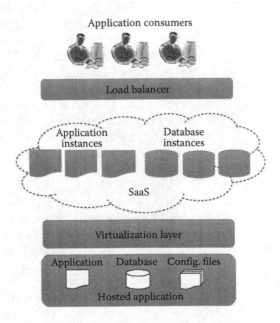

FIGURE 7.19
SaaS.

be more and unpredictable. The ability of handling the extra load decides the scalability of the application. The scalability of the application will be increased by the software load balancer, which will transfer the additional load to the new application/database server. Here, multiple application instances and database instances will be created to ensure high scalability. Some of the popular SaaS applications are Google Docs, Google Drive, and Microsoft Office 360. An overview of a SaaS application is illustrated in Figure 7.19.

Virtualization is used as an enabling technology to provide multitenant infrastructure, development platform, and SaaS. Additionally, there are other cloud services that use virtualization such as Network as a Service using network virtualization, Storage as a Service using storage virtualization, and Database as a Service using database virtualization.

7.6 Summary

Virtualization is a widely used technology in the IT industry to increase resource utilization and ROI. It allows the same physical infrastructure to be shared between multiple OSs and applications. The other benefits of virtualization include dynamic data center, green IT support, ease of administration, and improved disaster recovery. There are three types of approaches used to achieve virtualization, namely full virtualization, paravirtualization, and hardware-assisted virtualization. Full virtualization completely abstracts the guest OS from the underlying infrastructure. Paravirtualization provides partial abstraction of the guest OS from the underlying infrastructure with slight modification of the guest OS. In hardware-assisted virtualization, the hardware vendor itself offers the support for virtualization. Hypervisors are the key drivers in enabling virtualization in large-scale cloud data centers. There are two types of hypervisors available, namely type 1 or bare metal hypervisor and type 2 or hosted hypervisors. Type 1 hypervisors can directly interact with the underlying infrastructure without the help of the host OS. Type 2 hypervisors need the host OS to interact with the underlying infrastructure. Since hypervisors are used as the enabling technology in virtualized data centers, there are different types of attacks targeted on the hypervisors to disrupt the servers. Normally, the attacks are performed by malicious codes to compromise the hypervisors. The attacks may target both the guest OS or host OS. The attacks can be mitigated by strong guest isolation, frequent updates, enabling least privilege policies, monitoring tools, etc. Virtualization helps in creating multitenant cloud environment, where a single instance of the resource can be shared by multiple users. Cloud computing and virtualization are different. Cloud computing uses virtualization with utility computing to provide different services such as IaaS, PaaS, and SaaS.

Review Points

- *Virtualization* is a technology that changes the computing from physical infrastructure to logical infrastructure (see Section 7.1).

- *Processor virtualization* is the process of abstracting physical processor to the pool of virtual processor (see Section 7.2.1).

- *Memory virtualization* is the process of providing virtual main memory to the VMs that are abstracted from the physical main memory (see Section 7.2.2).

- *Storage virtualization* is a form of resource virtualization where a multiple physical storage is abstracted as a multiple logical storage (see Section 7.2.3).

- *Network virtualization* is the process of abstracting physical networking components to form a virtual network (see Section 7.2.4).

- *Data virtualization* aggregates the heterogeneous data from a different source to a single logical or virtual volume of data (see Section 7.2.5).

- *Application virtualization* allows the users to access the virtual instance of the centrally hosted application without installation (see Section 7.2.6).

- *Protection rings* are used to isolate the OS from untrusted user applications (see Section 7.3).

- *Full virtualization* is the process of completely abstracting the underlying physical infrastructure with binary translation and direct execution (see Section 7.3.1).

- *Paravirtualization* or OS-assisted virtualization partially abstracts the underlying infrastructure with hypercalls (see Section 7.3.2).

- *Hardware-assisted virtualization* eliminates the overhead of binary translation and hypercalls, where the hardware vendors itself support virtualization (see Section 7.3.3).

- *Hypervisor* or VMM is a software tool that enables virtualization (see Section 7.4).

- *Bare metal hypervisor* or type 1 hypervisor can run on physical infrastructure without any help from the host OS (see Section 7.4.1).

- *Hosted hypervisor* or type 2 hypervisors require the help of the host OS to communicate with the underlying infrastructure (see Section 7.4.1).

- *Hypervisor attacks* mostly target the VMM by the malicious code either from the guest OS or host OS (see Section 7.4.2).

- *Cloud computing* is different from virtualization by means of type of service, service delivery, elasticity, etc. (see Section 7.5).
- *IaaS* uses the processor, memory, storage, and network virtualization to provide the infrastructure services (see Section 7.5.1).
- *PaaS* virtualizes the development platform and provides it as a service to the developers (see Section 7.5.2).
- *SaaS* allows the multiple end users to share the single instance of centrally hosted software (see Section 7.5.3).

Review Questions

1. What is virtualization? List its benefits and drawbacks.
2. Explain how virtualization changes the computing in the IT industry.
3. Briefly explain how hardware resources such as processor, memory, storage, and networks can be virtualized.
4. Write short notes on data virtualization and application virtualization.
5. What are protection rings? Explain how it is used in virtualization.
6. Explain the different approaches used to achieve virtualization with a neat diagram.
7. Differentiate full virtualization, paravirtualization, and hardware-assisted virtualization techniques.
8. What is the role of hypervisor in virtualization? Briefly explain the different types of hypervisors with a neat diagram.
9. Differentiate type 1 and type 2 hypervisors.
10. Explain the different attacks targeted on hypervisors with a neat diagram.
11. Recommend some of the best practices to avoid/prevent the attacks on hypervisors.
12. Are virtualization and cloud computing the same? Justify your answer.
13. Explain how cloud computing is different from virtualization.
14. Compare and contrast cloud computing and virtualization.
15. Explain how virtualization is used as an enabling technology in delivering cloud services such as IaaS, PaaS and SaaS.

Further Reading

Alam, N. Survey on hypervisors. School of Informatics and Computing, Indiana University, Bloomington, IL.

Business and financial aspects of server virtualization. White Paper, AMD.

Carbone, M., W. Lee, and D. Zamboni. Taming virtualization. *IEEE Security and Privacy* 6(1): 65–67, 2008.

Chen, W. K. *Virtualization for Dummies*. Hoboken, NJ: Wiley Publishing, Inc., 2007.

Goth, G. Virtualization: Old technology offers huge new potential. *IEEE Distributed Systems Online* 8(2): 3, 2007.

Li, Y., W. Li, and C. Jiang. A survey of virtual machine system: Current technology and future trends. *2010 Third International Symposium on Electronic Commerce and Security (ISECS)*, IEEE Computer Society, 2010, pp. 332–336.

Mann, A. Virtualization 101: Technologies, benefits, and, challenges. White Paper, EMA Boulder, CO.

Mosharaf Kabir Chowdhury, N. M. and Boutaba, R. A survey of network virtualization. *Computer Networks* 54(5): 862–876, 2010. ISSN 1389-1286.

Perez-Botero, D., J. Szefer, and R. B. Lee. Characterizing hypervisor vulnerabilities in cloud computing servers. *Proceedings of the 2013 International Workshop on Security in Cloud Computing (Cloud Computing '13)*, ACM, New York, 2013, pp. 3–10.

Reuben, J. S. A survey on virtual machine security. Seminar of Network Security, Helsinki University of Technology, Helsinki, Finland, 2007.

Scarfone, K., M. Souppaya, and P. Hoffman, Guide to security for full virtualization technologies. NIST Special Publication, 800-125, 2011.

The benefits of virtualization for small and medium businesses. White Paper, VMware, Inc., Palo Alto, CA.

Uhlig, R., G. Neiger, D. Rodgers, A. L. Santoni, F. C. M. Martins, A. V. Anderson, S. M. Bennett, A. Kagi, F. H. Leung, and L. Smith. Intel virtualization technology. *Computer* 38(5): 48–56, 2005.

Understanding full virtualization, paravirtualization, and hardware assist. White Paper, VMware, Inc., Palo Alto, CA.

Virtualization in education. White Paper, IBM.

Virtualization is not the cloud. Technical article, Rackspace Support. Available [Online]: http://www.rackspace.com/knowledge_center/sites/default/files/whitepaper_pdf/Rackspace_Virtualization%20Is%20Not%20the%20Cloud_20120503.pdf (Accessed December 30, 2013).

Virtualization overview. White Paper, VMware, Inc., Palo Alto, CA.

8

Programming Models for Cloud Computing

Learning Objectives

The objectives of this chapter are to

- Give a brief description about the programming models available in cloud
- Point out the advantages and disadvantages of each programming model
- Point out the characteristic of each programming model
- Describe the working base of each programming model
- Point out the key features of the programming model
- Discuss the suitability of each programming model to different kinds of application

Preamble

Cloud computing is a broad area that has garnered the attention of several people and is now much more than a buzzword. This technology is one of the few that have directly affected human beings. In simple terms, a cloud offers services to its customers in several ways. One of the important ways is through Software as a Service (SaaS). In SaaS, a software application is given as a service to the users. There are several properties associated with *cloud* applications: scalability, elasticity, and multitenancy. Each and every cloud application should have the aforementioned properties.

This is what differentiates cloud from conventional applications. It is a known fact that any application involves a specific development strategy. In general, all development processes are characterized by the use of a specific programming model chosen based on the type of application we develop. The properties of application play a very important role in choosing a programming model. Thus, the development of this cloud application faces certain challenges.

Cloud computing is an emerging technology, and there are not many programming languages developed for it. This chapter broadly discusses the programming models available in cloud. Based on the approach or the method, the programming models can be classified into two. The first is using an existing programming model and extending it or modifying it to suit the cloud platform, and the second one is by developing a whole new model that will be suitable for cloud. These will be discussed in two separate sections. The first section of this chapter gives a brief introduction about programming models. It also discusses about cloud computing, its properties, and the impact that this field has on the technological world and the market. This section also discusses the differences between a usual application and a cloud application, as well as the big reason behind the quest for a new programming model. The existing programming models and languages used are discussed first, followed by the new programming models that are specifically designed for cloud. The chapter ends with a summary and some review questions.

8.1 Introduction

Programming model is a specific way or a method or approach that is followed for programming. There are several programming models available in general, each having their own advantages and disadvantages. Programming models form the base for any software or application development approach. The properties of these programming models decide their usage and their impact. As far as general programming models are concerned, these have continuously evolved. Each evolutionary step is characterized by certain changes to the existing methodology, and at each level, a new functionality is added, which facilitates the programmers in one or several ways. As the technology is growing, so are the problems. One programming model cannot be the solution for all the problems, even with a large number of advancements. In these cases, a whole new programming model is developed for a specific set of problems or for general use. To address an issue, any of the two approaches can be used.

Today, cloud computing is one of the most popular and extensively used technologies. As a technology, cloud is considered to be good and is popular because it allows users to primarily use computing, storage, and networking services on demand without spending for infrastructure (capital expenditure). An extensive research is going on in this segment. Many business organizations have started depending on cloud. The fast development of cloud was driven by its market-oriented nature.

Several big companies like Amazon, Google, and Microsoft have started using cloud extensively, and they intend to continue this trend.

With the advent of cloud computing, there came a new shift in the computing era. All the applications are being migrated to cloud as the future seems to be fully dependent on it. The service architecture of cloud has three basic components called service models, namely Infrastructure as a Service (IaaS), Platform as a Service (PaaS), and Software as a Service (SaaS). Among these three, SaaS is the most important model where software is given as a service from cloud. This software is specifically designed for cloud platform, and for designing a software or application, you have to use a specific programming model. Thus, a study about it is important.

Cloud has several properties that make it totally different from other conventional software applications, such as scalability, concurrency, multitenancy, and fault tolerance. There are generally two types of practices: one is extending the existing programming models or languages to cloud, and the other is to have a completely new model specifically designed for cloud. The programming models under these two areas are broadly discussed.

8.2 Extended Programming Models for Cloud

This section discusses the existing programming models that can be extended to a cloud platform. Not all programming models can be migrated to cloud. As discussed, there are several properties of cloud that make it a different technology from others. Thus, certain changes have to be made to address the issues of the cloud. Only certain programming models that have some similarity to cloud or have a structure that can eventually support certain parameters like scalability, concurrency, and multitenancy can be extended. Out of these properties, scalability and concurrency are considered to be very important for cloud. Scalability is the ability of the system to scale itself up or down according to the number of users. This is one of the reasons behind the popularity of cloud. Thus, this property should be taken care of by any programming model. Some of the programming models and frameworks that are suitable for cloud, including certain programming languages, are discussed in the following subsections.

8.2.1 MapReduce

MapReduce is a parallel programming model developed by Google [1]. This is one of the technologies that has a specific impact on the area of parallel computing as well as cloud computing. MapReduce uses a parallel programming paradigm for computation of a large amount of data by partitioning or segregating them into several pieces, processing them in parallel, and

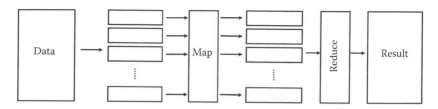

FIGURE 8.1
MapReduce.

then combining them to generate a single result. These processes are called map and reduce, respectively. The popular PaaS called Aneka also uses Map Reduce with .NET platform [2].

The first step in the map and reduce functions involves segregation of large data and sending it to map function. The map function performs map operation on data. Subsequently, the data are sent for reduce operation where the actual results are obtained. This is discussed in detail in the following subsections and shown in Figure 8.1.

8.2.1.1 Map Function

First, the large data are traced and segregated as key/value pairs. The map function accepts the key/value pair and returns an intermediate key/value pair. Usually, a map function works on a single document. The following formula depicts the exact operation:

$$Map(Key1,\ Val1) \rightarrow List(Key2,\ Val2)$$

The list of key/value pair obtained is subsequently sent to the reduce function.

8.2.1.2 Reduce Function

Once the list of key/value pair is prepared after the map functions, a sort and merge operation is applied on these key/value pairs. Usually, reduce function works on a single word. Input of distinct key/value pair is given and in return the group with the same key/value pair is obtained, as shown in the following equation:

$$Reduce(Key2, List(Val2)) \rightarrow List(Val3)$$

The resultant values give the final set of result.

The most popular example is counting the words in a file.

In the first step, the words (key) are assigned the values. The word represents the key, and the count represents the value. Whenever the word is encountered, the key/value pair is generated. This set of <word, count> pair obtained from the map function is sent to the reduce function. The result is sorted, and the reduce function combines all the words that have the same key and the *count* (value) is automatically incremented (added) for each pair obtained. Thus, the result gives each word with its count.

Key features

- Supports parallel programming
- Fast
- Can handle a large amount of data

8.2.2 CGL-MapReduce

CGL-MapReduce is another type of MapReduce runtime and was developed by Ekanayake et al. [3]. There are specific differences between CGL-MapReduce and MapReduce. Unlike MapReduce, CGL-MapReduce uses streaming instead of a file system for all communications. This eliminates the overheads associated with file communication, and the intermediate results from the map functions are directly sent. The application was primarily created and tested for a large amount of scientific data and was compared with MapReduce. Figure 8.2 shows the working architecture of CGL-MapReduce where its several components are depicted.

1. *Map worker*: A map worker is responsible for doing map operation.
2. *Reduce worker*: A reduce worker is responsible for doing reduce operation.

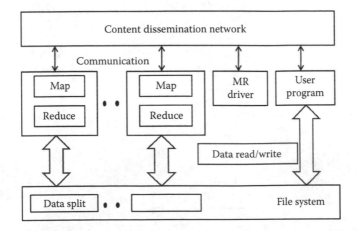

FIGURE 8.2
CGL-MapReduce. (Adapted from Ekanayake, J. et al., Mapreduce for data intensive scientific analyses, *IEEE Fourth International Conference on eScience'08, 2008 (eScience'08)*, Indianapolis, IN, IEEE, Washington, DC, 2008.)

3. *Content Dissemination Network*: Content dissemination network handles all the communication between the components using the NaradaBroker, which was developed by Ekanayake et al. [3].

4. *MRDriver*: MRDriver is a master worker and controls the other workers based on the instructions by the user program. CGL-MapReduce is different from MapReduce, the main difference being the avoidance of file system and usage of streaming. Several steps involved in CGL-MapReduce are depicted in Figure 8.3 and explained as follows [3]:

    a. *Initializing stage*: The first step involves starting the MapReduce worker nodes and configuration of the MapReduce task. Configuration of the MapReduce is a onetime process and multiple copies of it can be reused. This is one of the improvements of CGL-MapReduce, which facilitates efficient iterative MapReduce computations. These configured MapReduce are stored and executed upon the request of the user. Here, fixed data are given as input.

    b. *Map stage*: After the initialization step, MRDriver starts the map computation upon the instruction of the programmer. This is done by passing the variable data to the map tasks. This is

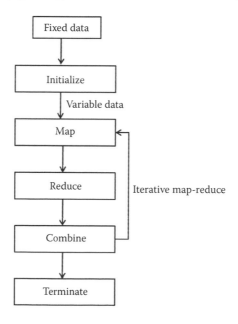

FIGURE 8.3
Steps involved in CGL-MapReduce. (Adapted from Ekanayake, J. et al., MapReduce for data intensive scientific analyses, *IEEE Fourth International Conference on eScience'08, 2008 (eScience'08)*, Indianapolis, IN, IEEE, Indianapolis, IN, 2008.)

relayed to workers for invoking configured map tasks. It also allows passing the results from one iteration to another. Finally, the map tasks are transferred directly to reduce workers using dissemination network.

c. *Reduce stage*: As soon as all the map tasks are completed, they are transferred to reduce workers, and these workers start executing tasks after they are initialized by the MRDriver. Output of the reduce function is directly sent to the user application.

d. *Combine stage*: In this stage, all the results obtained in the reduce stage are combined. There are two ways of doing it: if it is a single-pass MapReduce computation, then the results are directly combined, and if it is iterative operation, then appropriate combination is obtained such that the iteration continues successfully.

e. *Termination stage*: This is the final stage, and user program gives the command for termination. At this stage, all the workers are terminated [3].

Key features

- Uses streaming for communication
- Supports parallelization
- Iterative in nature
- Can handle a large amount of data

8.2.3 Cloud Haskell: Functional Programming

Cloud Haskell is based on the Haskell functional programming language. These functional programming languages are function based and work like mathematical functions. Usually, the output here is based on the input value only. Functional programming consists of components that are often immutable. Here, in Cloud Haskell, the immutable components form the basis. Immutable components are the components whose states cannot be modified after they have been created. Cloud Haskell is a domain-specific language [4].

The programming model is completely based on message-passing interface, which can be used for highly reliable real-time application. The main aim is to use concurrent programming paradigm. The Cloud consists of several applications and users who are independent and are large in numbers. Thus, to address the issue, concurrency has to be ensured for a large number of requests. Cloud Haskell provides the concurrency features with complete independence for the application. The data are completely isolated and the applications cannot breach the boundary and access other data. As mentioned earlier, the communication is completely based on message-passing model. According to the developers, there needs to be a separate cost model that would determine the data movement. There are several advantages of Haskell apart from the use of Erlang's fault tolerance model. Its main advantage is that it is a pure functional language. All the data are immutable by default and functions are idempotent. Idempotency here refers to the ability of the functions to restart from any points on its own without the use of any external entity such as a distributed database when the function is aborted due to hardware error.

A cost model has been introduced specifically in Cloud Haskell, which was not present in Haskell. Here, the cost model determines the communication cost between the processes. This model is not designed to have a shared memory. As the model supports concurrency and these processes are isolated, fault tolerance is expected to an extent and the failure of one process would not affect others. There are several other novel features that are included into this model such as serialization. Whenever a programmer tries to use the distributed code, he had to run the code in a remote system. This was not possible in Haskell versions. Thus, in this model, the developers introduce a mechanism that will automatically take care of this issue without extending the compiler.

In simple terms, the serialization is taken care of with full abstraction so that the programmer does not know it. The model also takes care of failure, which is an important issue in cloud. There are many ways in which failure might affect the functioning of the system. As the number of systems connected and users increase, failure rate also increases. Usually, to confront partial failure, the whole system is restarted. This is considered as one of the inefficient solutions as when it comes to cloud the number of distributed nodes is high, and so restarting a system can prove to be very costly. Thus, to maintain fault tolerance, Erlang's method is used. Even if a function fails because of certain reasons, it can be separately restarted without affecting the other parts and processes. Here, the components are immutable and otherwise called pure functions [4].

Key features

- Fault tolerant
- Domain-specific language
- Uses concurrent programming paradigm
- No shared memory
- Idempotent
- Purely functional

8.2.4 MultiMLton: Functional Programming

MultiMLton is specifically used for multiprocessor environments and is an extension of MLton. MLton is an open-source, whole-program, optimizing standard ML compiler [5]. It is used for expressing and implementing various kinds of fine-grained parallelism. It has the capability to efficiently handle a large number of lightweight threads. It also combines new language abstractions and compiler analyses. There are several differences between MultiMLton and its other counterparts like Erlang and ConcurrentML. It provides additional parallelism whenever possible and whenever the result is profitable by using deterministic concurrency within threads. This increases speed and will eventually give good performance. It also provides message-passing aware composable speculative actions [5]. It provides isolation among groups of communicating threads [5]. The communication between the threads happens through simple message-passing techniques. It allows construction of asynchronous events that integrate abstract synchronous communication protocols.

The MultiMLton model consists of a component that is known as a parasite. These parasites are mini threads that depend on a main or master thread for its execution. As mentioned earlier, communication between threads can be synchronous and asynchronous. Figure 8.4 shows a simple communication between the parasite of one thread and another thread.

Key features

- Can efficiently handle a large number of lightweight threads.
- Suitable for multiprocessor environment.
- Deterministic concurrency is used wherever necessary.

8.2.5 Erlang: Functional Programming

Erlang is one of the important types of functional programming language. It was developed by Joe Armstrong and is one of the very few languages that are highly fault tolerant. Even Haskell (another programming language) uses Erlang's fault tolerance model [4].

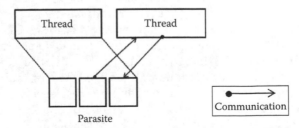

Parasite

FIGURE 8.4
Message passing.

Erlang is a concurrent programming language that uses asynchronous message-passing techniques for communication. Here, communication can be done only through a message-passing mechanism. Creating and deleting processes are very easy, and message-passing processors require less memory; hence, it is considered as lightweight. It follows a process based on a model of concurrency.

This programming model is used to develop real-time systems where timing is very critical. Fault tolerance is one of the important properties of Erlang. It does not have a shared memory as mentioned earlier; all the interactions are done through message passing. It has good memory management system and allocates and deallocates memory automatically with dynamic (runtime) garbage collection. Thus, a programmer need not worry about the memory management, and thereby memory-related errors can be avoided. One of the most striking properties of Erlang is that it allows code replacement in the runtime. A user can change the code while a program is running and can use the same new code simultaneously.

Erlang, by definition, is a declarative language. It can also be regarded as a functional programming language. Functional programming languages are the languages that depend only on input data. This property is usually called immutability. This is the base property that makes functional programming the most suitable programming model for cloud. These properties allow the programmers to use concurrency and to scale the system, which is one of the important properties of cloud.

Key features

- Fault tolerant
- Robust
- Concurrent
- Uses functional programming model

8.2.5.1 CloudI

CloudI is an open-source cloud computing framework developed by Michael Truog. It allows building a cloud platform (a private cloud) without actual virtualization. It is basically for back-end server processing tasks. These tasks require soft real-time transaction processing external to database usage in any language such as C, C++, Python, Java, Erlang, and Ruby [6]. CloudI is built over the Erlang programming language. It is aimed at being fault tolerant and has a highly scalable architecture (a cloud at the lowest level; bringing Erlang's fault tolerance to polyglot development). It consists of a messaging application programming interface (API) that allows CloudI services to send requests.

CloudI API supports both publish/supply and request/reply messages. Apart from fault tolerance, which is one of its important properties, CloudI can easily support other languages or a polyglot environment. It is not necessary for a user

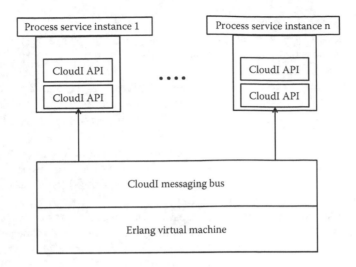

FIGURE 8.5
CloudI communication. (Adapted from Bringing Erlang's fault-tolerance to polyglot development, available [online]: http://www.toptal.com/erlang/a-cloud-at-the-lowest-level-built-in-erlang.)

to know Erlang completely to work with CloudI. Any service module in CloudI is a miniature of the central module. Each service module contains CloudI API.

Here, all the communications is done through CloudI API. Figure 8.5 depicts the interrelations and the communication between the CloudI service instance and process service instances. The CloudI messaging bus acts as an interface between the APIs and the Erlang VM. Each process service consists of CloudI API and is responsible for communication.

Key features

- Uses functional programming model
- Fault tolerant
- Supports many languages
- Supports extensive scalability

8.2.6 SORCER: Object-Oriented Programming

Service-ORiented Computing EnviRonment (SORCER) is an object oriented cloud platform for transdisciplinary service abstractions proposed by Michael Sobolewski, which also supports high-performance computing. It is a federated service-to-service (S2S) metacomputing environment that treats service providers as network peers with well-defined semantics of a service object-oriented architecture [8].

Transdisciplinary computing consists of several service providers. All service providers have a collaborative federation, and they use an orchestrated workflow for their services. As soon as the work is over, the collaboration is dissolved and the service providers search for other federations to join. This method is wholly network based. Here, each service is an independent and self-sustaining entity called remote service provider, which performs specific network tasks. Thus, the whole method is network centric. For controlling the whole system, an operating system (OS) is required, which will enable the providers to give instructions related to interaction between services.

The users do not use this technology as for managing the orchestration and the interaction between the services. The users have to know command line codes and script, thereby making it a tedious job. Hence, to solve this problem, service-oriented operating system (SOOS) is used. Here, metaprogram is considered as a process expression of remote component programs, and the SOOS makes all the decisions about place, time, and other details regarding running the remote component program. The metaprograms are programs that remotely manipulate the other programs as its data. Here, it is used as a specification for service collaboration.

Figure 8.6 shows the difference between the different types of computing mechanisms. These include multidisciplinary, interdisciplinary, and transdisciplinary computing. The metaprogram is used for transdisciplinary computing. The diagram clearly shows the place where metaprogram is executed. Figure 8.6a shows a discipline that consists of several data, operations, and control components connected together in a workflow to give the result. Figure 8.6b shows multidisciplinary models that involved two or more disciplines but not intercommunication between the disciplines. Similarly, Figure 8.6c shows an interdisciplinary model where intercommunication between the disciplines is involved. Finally, Figure 8.6d shows the transdisciplinary models where multiple disciplines with interdisciplinary communication are present, and here the components from outside the discipline can also contribute to the results. These transdisciplinary models use the metaprogram.

The whole concept of SORCER is based on remote processing. Here, the corresponding computational components are autonomically provisioned by using metainstructions that are produced by service metaprocessor. This is done instead of sending the executable files directly through the Internet. Metaprogram and metainstruction are service command and service instruction, respectively.

Here, a metacompute OS is used, which manages the whole dynamic federation of service providers and resources. With its own control strategy, it allows all the service providers to collaborate among themselves, which is done according to the metaprogram definition given. The metaprogram, in terms of metainstruction, can be submitted to metacompute OS.

There are several layers in SORCER, as shown in Figure 8.7, and among them SORCER OS is the most important. In the figure, the layers are grouped according to their functionality. The command virtual platform and the object virtual platform collectively form the platform cloud.

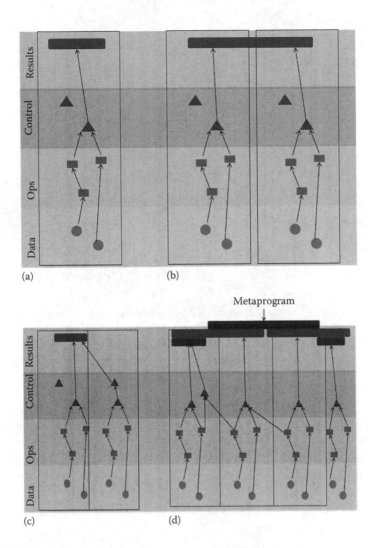

FIGURE 8.6
Types of operations: (a) discipline, (b) multidisciplinary, (c) interdisciplinary, and (d) transdisciplinary. (Adapted from Sobolewski, M., Object-oriented service clouds for transdisciplinary computing, in *Cloud Computing and Services Science*, Springer, New York, 2012, pp. 3–31.)

The service nodes and the domain specific (DS) service providers collectively do the service cloud operation. The aforementioned operations with command native platforms and evaluator filters are controlled by metaprocessor. The whole thing is managed by service operating system and above which service programs or service requestors request the services.

SORCER is focused on metamogramming. Metamogramming is a process in which metamodeling or programming environment is used for creation of symbolic process expression. Thus, SORCER introduces a

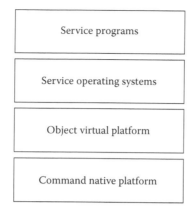

FIGURE 8.7
SORCER architecture. (Adapted from Sobolewski, M., Object-oriented service clouds for transdisciplinary computing, in *Cloud Computing and Services Science,* Springer, New York, 2012, pp. 3–31.)

metacompute OS that consists of all the required system services, which includes object metaprogramming, federated file system, and autonomic resource management. These properties are to support service object–oriented programming. It basically provides solution for multiple transdisciplinary applications that require complex transdisciplinary solutions. SORCER is deployed over FIPER.

FIPER is federated intelligent project environment, and its main aim is to provide federation of distributed service objects that provide engineering data, applications, and tools on a network. In short, SORCER metamodeling architecture is based on domain/management/carrier, or the DMC triplet [8].

Key features

- Object-oriented nature
- Service-oriented architecture
- Transdisciplinary

8.2.7 Programming Models in Aneka

Aneka is a PaaS developed by Manjrasoft, Inc. There are three programming models defined and used by Aneka, which are discussed in the following.

8.2.7.1 Task Execution Model

In this model, the application is divided into several tasks. Each task is executed independently. As defined by the developers of Aneka, tasks are work units that are executed by the scheduler in any order. As it involves a large number of tasks, it is very much suitable for distributed applications, specifically applications where several independent results are involved, and these independent results are then combined to give a final result. For these kinds of problems, the independent results can be considered as output from the tasks, and later these results can be combined by the user. It also supports dynamic creation of tasks [9].

Key features

- Suitable for distributed application
- Independent tasks

8.2.7.2 *Thread Execution Model*

In thread model, the applications are executed using processes. Each process consists of one or more threads. As defined by the developers of Aneka, threads are a sequence of instructions that can be executed in parallel with other instructions. Thus, in thread execution model, the execution is taken care by the system [10].

Key features

- Supports heavy parallelization.
- Multiple thread programming is available.

8.2.7.3 *Map Reduce Model*

This model is similar to the actual MapReduce model described in the first section. The Map Reduce model is implemented in the Aneka platform.

8.3 New Programming Models Proposed for Cloud

This section discusses about the programming models that are newly designed for cloud. These programming models are designed from scratch. As these models are specifically designed for cloud, the important properties of cloud, such as scalability, distributed nature, and multitenancy, form the base for designing algorithm. Unlike the previous method, there is no necessity to go according to any language specification. Mainly all the approaches concentrate on the distributed nature of algorithms. The concept of a new programming model came because of certain disadvantages that were found in existing programming models. The main disadvantage is that you cannot eliminate certain basic properties of existing approaches. Only new parameters can be added or slight modification can be done so as to achieve the required goal. Whereas when a new programming model is designed, the designer has full liberty to include anything without any restriction. The following section also includes a few APIs and toolkits that are specifically designed for cloud.

8.3.1 Orleans

Orleans is a framework developed by Microsoft for creating cloud applications [11]. It is an attempt by Microsoft to give a software framework that will facilitate both client and server sides of cloud application development. This uses a concurrent programming paradigm.

As this application is designed for cloud, there are several characteristics that are considered. The important features of Orleans include

- Concurrency
- Scalability
- Reduced errors
- Security

The basic idea around which this whole framework is developed is the use of grains. These grains are logical computational units that are considered to be the building blocks of Orleans. The grains are independent and isolated, and all the processes are executed by using grains. A grain can have several activations. These activations are instantiations of grains in the physical server. Activations are helpful as several processes can concurrently be executed by using several activations. Though the model supports concurrency by default, a grain does not execute processes concurrently. A grain can serve only one request at a time. The next request can be served only once a request is completed.

There can be zero, one, or several activations. Thus, to support parallel execution, several activations of a single grain can be created and several requests can be served in parallel. Several advantages of this feature, as quoted by the developers, include increasing throughput and reducing latency. This feature has a great impact on Orleans. They communicate with each other by passing messages. Each grain has a state. The state determines the current position of a grain. When several activations are created for a single grain, there is a possibility that the state of the grain might change. This might lead to inconsistencies among all the activations. This issue is also addressed by Orleans.

Scalability is another issue that needs to be considered when it comes to the development of cloud application. Orleans on its part considers scalability as an important issue and tries to provide a solution. The concept of grains and activation itself considers scalability, and in addition to that, Orleans addresses the issue by the use of sharded database. Sharded database is horizontally partitioned into several components with each component (shard) consisting of several rows. This sharded database architecture is used in the back end to support the grains. Thus, grains can be fully independent and can have an independent computation. This is called grain domain. The property of distributed and independent computation is very effective since all the applications depend on database. If there is a single database, then it becomes a bottleneck. Thus, everything would depend on the single database, and if that database fails, the whole system collapses. But when the system is distributed in nature, it is not the case.

The failure of a single application results in a failure of a part of a database that would not affect other chunks of databases and thereby would not affect other applications. Thus, in a way, it becomes fault tolerant and is extensively usable and scalable as application isolation and independence is guaranteed. There is no restriction in creation of number of grains and number of activations. Thus, as the number of request increases, so are the number of processes, then we can simultaneously increase the grains and activations to serve those requests. These processes will not affect each other as they are completely isolated. And so, the concept of grains and activation itself is the solution for scalability. Figure 8.8 depicts the layer at which the Orleans works. The Orleans is implemented as a library in C# over .Net framework [11].

Apart from these issues, the model introduces transactions, and according to the developers there are several reasons for using transactions. The first reason is that the activations can change the state of the grains; thus, all the changes made by the processes need to be maintained. To maintain consistency, there is a necessity to have a separate mechanism for maintaining it as several activations can change the data simultaneously. The other reason is replication of data. Replication of data, which is inevitable in the case of cloud technology, has to be done carefully. This is because several applications are dependent on data, and replication of data can

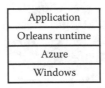

FIGURE 8.8
Orleans system. (Adapted from ORLEANS: A framework for cloud computing, available: http://research.microsoft.com/en-us/projects/orleans/.)

result in inconsistency. Thus, transactions are introduced. Apart from scalability, Orleans introduces restricted concurrency [11].

Key features

- Fault tolerant
- Sharded database
- Efficient transaction handling
- Restricted concurrency
- Security

8.3.2 BOOM and Bloom

Berkeley orders of magnitude (BOOM) is a project undertaken by UC Berkeley. It aims at creation of a programming framework designed for cloud. Bloom is a programming language created for the project BOOM. Bloom programming language is extensively distributed in nature. BOOM is highly distributed and unstructured in nature. As the name suggests, this was designed so that people can design large systems as cloud. It tries to follow a nontraditional approach to an extent possible. The approach developed is disorderly and data centric. BOOM aims at less code and more scalability. It uses a fully distributed approach. BOOM uses BOOM FS, which is a Hadoop distributed file system (HDFS). According to the analytics, the BOOM FS is 20% better than HDFS. The design goals of Bloom are as follows [13]:

- Familiar syntax
- Integrated with imperative languages
- Modularity, encapsulation, and composition

BLOOM uses the CALM principle and is built by using dedalus. The CALM principle is used for building automatic program analysis and visualization tools for reasoning about coordination and consistency, whereas dedalus is a temporal logic language. Dedalus is a time-oriented distributed programming language. It focuses on data-oriented nature based on time rather than space. Data related to space is not extensively used. It is not necessary for the programmer to understand the behavior of the interpreter or compiler since Dedalus is a pure temporal language [14].

Key features

- Modularity, encapsulation, and composition
- Extensively distributable
- Unstructured

8.3.3 GridBatch

GridBatch is a system that allows programmers to easily convert a high-level design into the actual parallel implementation in a cloud computing–like infrastructure. This was developed by Liu and Orban [15]. It does not parallelize automatically, and a programmer has to do this after understanding the application and should be able to decide upon its parallelizability factors. GridBatch provides several operators facilitating the programmers to do the tasks for parallelization. Usually, to implement an application, it should be broken down into 296 pieces of smaller tasks and the partition should be such that it has a maximum performance. To facilitate these things, GridBatch provides a set of libraries. These libraries have everything in it. The programmer needs to know how to use the library, and by using the library and operators effectively, a programmer can achieve the highest efficiency. GridBatch is specifically designed for cloud applications, but because of its simple parallel programming nature, it can be effectively used for grid computing and cluster computing as well. It supports the aforementioned environments better than its other counterparts, and it gives higher programmer productivity. As far as GridBatch is concerned, an application should have a large amount of data parallelism to have large performance gains. This GridBatch framework is suitable for analytical applications that involve a large amount of statistical data [15].

Key features

- Simple
- Supports efficient parallelization
- Better performance gains

8.3.4 Simple API for Grid Applications

Simple API for grid applications (SAGA) is a high-level API that provides a simple, standard, and uniform interface for the most commonly required distributed functionality. It was developed by Goodale et al. [16] and is mainly used for distributed applications. It also provides separate toolkits to work with distributed applications and provides assistance for implementing commonly used programs. The SAGA API is written in C++ with C and Java language support. A library called engine provides support for runtime environment decision making by using a relevant adapter, and this is the main and most important library [17]. SAGA uses task model and asynchronous notification mechanism for defining a concurrent programming model.

Here, the concurrent units are allowed to share object states, which are one of the necessary properties. This property creates an additional overhead of concurrency control. Thus, enforced concurrency mechanism needs to be adopted. But, enforced concurrency mechanism can be a problem in some cases as sometimes it might not suit a certain problem domain and might result in unnecessary increase in complexity and overhead. Thus, concurrency control mechanisms are not enforced. The user has to take care of concurrency. SAGA works directly over the grid toolkits like global and condor that are used for creating grids. Unlike open grid service architecture (OSGA), SAGA is not a middleware; instead, SAGA can work over OSGA. The distributed applications are deployed over SAGA.

The applications are works over distributed application pattern or user modes. Though SAGA works directly over toolkits, it is not a middleware. It is an API as the name suggests. Figure 8.9 depicts the architecture of SAGA with all the layers involved in an application development. Here, the SAGA API works over several grid components such as condor and globus. Several types of distributed application managing modes are deployed over SAGA. These involve process like MapReduce and process for managing hierarchical jobs. Over this, the actual distributed application works. An application can work directly either over SAGA or over distributed application pattern layer. Usually, the intermediate layer between SAGA and application facilitates an efficient management of resources.

Further, SAGA can be elaborated based on the two types of packages available. The first one is the packages that are related to look and feel. These look and feel API packages are important because it is responsible for providing the *look* and *feel* throughout. These include task model, security, monitoring, error handling, and attributes. Another package is called SAGA API packages. These are the basic packages that are required for SAGA.

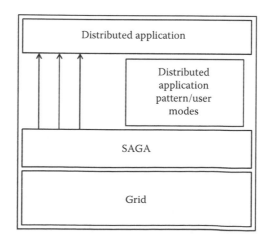

FIGURE 8.9
SAGA architecture. (Adapted from SAGA tutorial, available: http://www.cybertools.loni.org/rmodules.php.)

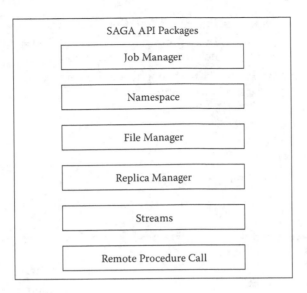

FIGURE 8.10
SAGA API packages.

These include job management, namespace, file management, replica management streams, and remote procedure call. Figure 8.10 depicts these packages.

1. *Job management*: As SAGA involves a large number of jobs to be submitted to the grid, job management function is used. This is used for submission, control, and monitoring the jobs. The function of the job manager is to submit the job to the grid resource, describe the procedure for controlling these jobs, and retrieve information about running and completed jobs. The submission of jobs is done for two modes: batch and interactive.

2. *Namespace management*: Namespace management is responsible for managing namespaces, and it works collectively with file management package.

3. *File management*: The files that are part of this SAGA are managed by the application. These files are accessed by the SAGA application without knowing the location of the file. This, along with namespace, is responsible for several operations on the file like read, seek, and write.

4. *Replica management*: Replica management describes interactions with replica systems [17]. A replica is a file registered on a logical file. This is used for creation and maintenance of logical file replicas and to search a logical file on metadata [17].

5. *Streams management*: The main function of a stream manager is to apply the simplest possible authenticated socket connection with hooks to support application-level authorization.

6. *Remote procedure call*: This manages all the remote procedure calls and provides methods for communication.

Key features

- Simple API
- C and Java language support

8.4 Summary

The users using cloud are increasing rapidly; thus, the demand for cloud is very high. The way an application is developed in the usual environment and the way it is developed in cloud are different. This chapter discusses about the programming models that are proposed for cloud computing. As cloud computing involves development of several applications, there needs to be a specific programming model that will facilitate the development of cloud application. As cloud applications are different, the existing programming model cannot be used as such. This chapter discusses both the types of programming models for cloud: the existing programming models that can be extended to cloud and the new programming models proposed. In the first category, functional programming, parallel programming, and object-oriented programming are extended to cloud. Existing programming languages that suit cloud are also discussed. Similarly, as far as new models are concerned, the distributed nature of cloud is mainly taken into account. Several new programming models proposed are discussed as well as a few frameworks specifically designed for cloud programming.

Review Points

- *Sharded database*: Sharded database is horizontally partitioned into several components with each component (shard) consisting of several rows (see Section 8.3.1).
- *Grain*: Logical computational units considered as building blocks of Orleans (see Section 8.3.1).

- *Parasite*: A parasite is a small (lightweight) thread that depends on other threads (main) (see Section 8.2.4).
- *Immutable components*: These are the components whose state cannot be modified after they have been created (see Section 8.2.3).
- *Idempotency*: It is the ability of a function to restart without the use of any external entity or recovery mechanism, even after the system has encountered a hardware failure (see Section 8.2.3).
- *Content dissemination network*: Content dissemination network is responsible for managing communication between different components of CGL-MapReduce (see Section 8.2.2).
- *Metamogramming*: A process in which metamodeling or programming environment is used for creation of symbolic process expression (see Section 8.2.6).
- *Metaprogram*: It is a program that manipulates other programs remotely as its data (see Section 8.2.6).
- *GridBatch*: It is a system that allows programmers to easily convert a high-level design into the actual parallel implementation in a cloud computing–like infrastructure (see Section 8.3.3).
- *Condor and globus*: Grid toolkits used for creating grids (see Section 8.3.4).
- *Fault tolerant*: It refers to the ability of the system to withstand error (see Section 8.2.5).

Review Questions

1. What is the main difference(s) between Cloud Haskell and Erlang?
2. What are the pros and cons of having a new programming model?
3. Which programming language is highly fault tolerant and for which type of application was it originally designed?
4. Point out the difference(s) between MapReduce and CGL-MapReduce.
5. In Orleans, failure of one portion of the application would not affect the whole application. Justify.
6. Does Cloud Haskell have its own fault-tolerance mechanism? If no, then on whose fault mechanism it is based?
7. Which programming model (language) allows runtime code replacement?
8. What are the main reasons for using transactions in Orleans?

9. What are the components of SAGA?

10. Does GridBatch parallelize the code automatically? If yes, then how? Else, who is responsible for parallelization of the code?

References

1. Dean, J. and S. Ghemawat. Mapreduce: Simplified data processing on large clusters. *ACM Communications*, 51: 107–113, January 2008.
2. Jin, C. and R. Buyya. Mapreduce programming model for. Net-based cloud computing. *Euro-Par 2009 Parallel Processing*, Delft, The Netherlands. Springer, Berlin, Germany, 2009, pp. 417–428.
3. Ekanayake, J., Pallickara, S., and G. Fox. Mapreduce for data intensive scientific analyses. *IEEE Fourth International Conference on eScience*, Indianapolis, IN. IEEE, Washington, DC, 2008.
4. Epstein, J., A. P. Black, and S. Peyton-Jones. Towards Haskell in the cloud. *ACM SIGPLAN Notices* 46(12): 118–129, 2011.
5. Sivaramakrishnan, K. C., L. Ziarek, and S. Jagannathan. CML: Migrating MultiMLton to the cloud.
6. A cloud at the lowest level. Available [Online]: http://cloudi.org/. Accessed November 10, 2013.
7. Bringing Erlang's fault-tolerance to polyglot development. Available [Online]: http://www.toptal.com/erlang/a-cloud-at-the-lowest-level-built-in-erlang. Accessed December 10, 2013.
8. Sobolewski, M. Object-oriented service clouds for transdisciplinary computing. In *Cloud Computing and Services Science*. Springer, New York, 2012, pp. 3–31.
9. Developing thread model applications—ManjraSoft. Available [Online]: www.manjrasoft.com/download/ThreadModel.pdf. Accessed November 5, 2013.
10. Developing task model applications—ManjraSoft. Available [Online]: www.manjrasoft.com/download/TaskModel.pdf. Accessed November 6, 2013.
11. Bykov, S. et al. Orleans: A framework for cloud computing. Technical Report MSR-TR-2010-159, Microsoft Research, 2010.
12. ORLEANS: A framework for cloud computing. Available [Online]: http://research.microsoft.com/en-us/projects/orleans/. Accessed November 10, 2013.
13. Neil, C. Cloud programming: From doom and gloom to BOOM and Bloom. UC Berkeley, Berkeley, CA.
14. BOOM: Berkeley Orders of Magnitude. Available [Online]: http://boom.cs.berkeley.edu/. Accessed November 3, 2013.
15. Liu, H. and D. Orban. Gridbatch: Cloud computing for large-scale data-intensive batch applications. *8th IEEE International Symposium on Cluster Computing and the Grid, 2008 (CCGRID'08)*. IEEE, 2008.
16. Tom, G. et al. A simple API for grid application (SAGA). Available [Online]: www.ogf.org/documents/GFD.90.pdf. Accessed November 20, 2013.

17. Miceli, C. et al. Programming abstractions for data intensive computing on clouds and grids. *Proceedings of the 2009 9th IEEE/ACM International Symposium on Cluster Computing and the Grid.* IEEE Computer Society, 2009.

18. SAGA Tutorial. Available [Online]: http://www.cybertools.loni.org/ rmodules.php. Accessed December 5, 2013.

Further Reading

Developing MapReduce.NET applications—ManjraSoft. Available [Online]: www. manjrasoft.com/download/2.0/MapReduceModel.pdf.

Erlang programming language. Available [Online]: http://www.erlang.org/.

9

Software Development in Cloud

Learning Objectives

The main objective of this chapter is to introduce the concept of Software as a Service (SaaS) and its development using Platform-as-a-Service (PaaS) technology. After reading this chapter, you will

- Understand how SaaS applications are different from traditional software/application
- Understand how SaaS benefits the service providers and the end users
- Understand the pros and cons of different SaaS delivery models
- Understand the challenges that are introduced by SaaS applications
- Understand how to develop a cloud-aware SaaS applications using PaaS technology

Preamble

This chapter gives an insight about SaaS applications that are different and offer many advantages when compared to traditional applications. We cannot choose the SaaS delivery option for all kinds of applications. The suitability of SaaS is also discussed in this chapter as well as the many SaaS deployment and delivery models available for SaaS development. Even though SaaS applications offer more advantages to the consumers, they bring a lot of challenges to developers with regard to SaaS application development, and this chapter will further discuss these challenges. An overview of other cloud service models such as Infrastructure as a Service (IaaS) and Platform as a Service (PaaS) that can be leveraged to develop multitenant-aware, scalable, and highly available SaaS applications is detailed in this chapter. This chapter also gives an idea about achieving secured multitenancy at the database level with different multitenancy models. Finally, the chapter discusses about the monitoring and SLA maintenance of SaaS applications.

9.1 Introduction

SaaS is a promising software delivery and business model in the information technology (IT) industry. The applications will be deployed by the SaaS provider in their managed or hosted infrastructure. The end users can access the hosted application as a service whenever there is a need. For using SaaS, the end users need to pay the full license fee. They can pay for what they used or consumed. To access the SaaS application, the customers need not install the software on their devices. It can be accessed from the service provider infrastructure using a simple web browser over the Internet. In the traditional software delivery model, the relationship between the end user and the software is one to one and licensing based. The end users have to buy the software from the vendors by paying a huge licensing amount. Some heavyweight applications need a high computing power. So, the end users have to buy the required hardware also. So, the initial investment to use the software is high, and if you look at the usage, it will be very low. To overcome this disadvantage of the traditional software delivery model, companies started developing the SaaS application, which has one-to-many relationships with the end users and the software. SaaS follows the multitenant architecture, and it allows many customers to share a single instance of the software, popularly known as multitenancy. Because of its cost-effective nature, many customers started moving to SaaS applications rather than the traditional licensing-based software. SaaS applications benefit not only the end users but also the service providers. In the traditional application service provider (ASP) model, the service providers host the applications on their own data center to benefit the end user. As the application delivery is one to one, the ASPs manage dedicated infrastructure for each customers, which forces them to invest a huge amount in managing the infrastructure, reducing their return on investment (ROI). Then the ASPs and the independent software vendors (ISVs) started using the alternate software delivery model (SaaS), which increased their ROI and at the same time benefitted the users. Now, most of the traditional ASPs and ISVs started realizing the business benefits of SaaS and started the SaaS development business.

9.1.1 SaaS Is Different from Traditional Software

The SaaS delivery model is different from the traditional license-based traditional software. The following discusses the many characteristics that differentiate the SaaS application from traditional applications:

- SaaS provides web access to commercial software on pay-as-you-use basis.
- SaaS applications are developed, deployed, and managed from a central location by the service provider.
- It allows the same instance of the SaaS application to be shared by multiple customers or tenants.
- Updates are performed by the service provider, not by the user.
- SaaS applications allow service integration with other third-party services through application programming interfaces (APIs) provided by them.

9.1.2 SaaS Benefits

SaaS applications provide cost-based benefits to the customers. It is also an on-demand, easy, and affordable way to use the application without a need to buy it. Additionally, SaaS solutions are easy to adopt and integrate with other software. Some of the notable benefits of SaaS are mentioned in the following:

1. *Pay per use*: SaaS applications are consumed by the end users on a pay-per-use basis. SaaS applications are on a subscription basis and allow the customers to use and disconnect the service as they wish. In the traditional software delivery model, the customers need to pay the full amount even if they use it very less.

2. *Zero infrastructure*: For installing and running traditional software, customers need to buy the required hardware and software, increasing the capital expenditure. In SaaS customers, there is no need to buy and maintain the infrastructure, operating system, development platforms, and software updates.

3. *Ease of access*: The SaaS application requires a simple web browser and an Internet connection to access it. The template-based responsive user interface (UI) will adapt automatically to the end user device, increasing the user experience and ease of access.

4. *Automated updates*: In the traditional software delivery model, the customers need to perform the bulk update, which is an overhead. But in SaaS, the service provider will perform the automated

updates. So, the customers can access the most recent version of the application without any updates from their side.

5. *Composite services*: Using SaaS applications, we can integrate other required third-party web services or cloud services through their API. It allows us to create a composite service.

6. *Dynamic scaling*: SaaS applications are used by a diverse user community. The load on the application will be dynamic and unpredictable. But with the dynamic load balancing capability, SaaS applications can handle any additional loads effectively without affecting their normal behavior.

7. *Green IT solutions*: SaaS applications are supporting the Green IT solutions. Since SaaS applications are multitenant and share the same resources and application instances, buying additional hardware and resources can be eliminated. The high resource utilization allows the application to consume less energy and computing power. SaaS solutions are becoming smarter and have energy-aware features in it that does not consume much resource for its operation.

9.1.3 Suitability of SaaS

SaaS applications are used by many individuals and organizations for its cost-effective nature. Their adoption by large enterprises also increases in a fair amount. But we cannot use SaaS applications in all places. The following are some applications where SaaS may not be the best option:

- Some real-time applications where fast processing of data is needed
- Applications where the organization's data are more confidential and the organization does not want to host their data externally
- Applications where existing on-premise applications fulfill the organization's needs

The following are examples where SaaS is the best option:

- For applications where the end user is looking for on-demand software rather than full-term/licensing-based software
- For a start-up company that cannot invest more money on buying licensed software
- For applications that need the accessibility from handheld devices or thin clients
- For applications with unpredictable and dynamic load

9.2 Different Perspectives on SaaS Development

The SaaS model provides web access to the commercial software. Since SaaS applications are deployed and managed by the service providers, customers are getting access to software services without any overhead of maintaining underlying infrastructure and platform. SaaS providers also can reduce the maintenance overhead by choosing other appropriate cloud services such as IaaS and PaaS, reducing the capital and operation expenditure of maintaining the servers. SaaS can be deployed and delivered from the traditional infrastructure or cloud infrastructure. There are different SaaS deployment and delivery models available to benefit the service provider and customer, which is discussed in this subsection.

9.2.1 SaaS from Managed Infrastructure and Platform

This model uses the traditional infrastructure and the platform for developing and deploying the SaaS application. Cloud computing characteristics will be satisfied only at the SaaS layer. In the other two layers (platform and infrastructure), the cloud characteristics will not be satisfied. This means that the underlying infrastructure and platform are not cloud enabled. The degree of multitenancy is also very low in this type of model as multitenancy is not achieved at the PaaS and IaaS levels. The developed SaaS application will be delivered to the customers in a one-to-many model. Figure 9.1 illustrates the concept of delivering SaaS from the self-managed infrastructure and platform.

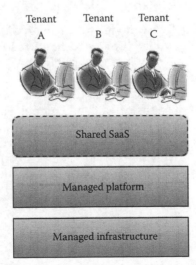

FIGURE 9.1
SaaS delivery from managed infrastructure and platform.

Pros

- This type of SaaS delivery model ensures more security to the user data as the infrastructure and platform are maintained by the SaaS provider.
- The SaaS provider gets full control over the infrastructure and development platform.
- There is no problem of vendor lock-in. The application can be easily migrated to any other infrastructure without any major modification.

Cons

- More overhead in maintaining the underlying infrastructure and platform.
- The service providers have to invest more on the infrastructure. So, this model is not suitable for SaaS development companies that do not have much amount to invest.
- Since the development environment is managed by the service provider, there is an additional overhead of maintaining the scalability and availability of the application.
- Resource utilization will be very low.

9.2.2 SaaS from IaaS and Managed Platform

In this type of service delivery model, the SaaS providers can use the infrastructure provided by any IaaS provider. The infrastructure provider may be a public or private IaaS provider. The public infrastructure will be chosen if the SaaS application does not require more security to the data. If the application needs more security and at the same time more resource utilization, they can choose the private IaaS model. Here, the multitenancy will be achieved at the infrastructure and application layers. The service providers have to manage their own development platform. Since the infrastructure is given by a third party, there is a possibility of vendor lock-in at the IaaS layer. Figure 9.2 illustrates SaaS delivery from the IaaS and self-managed platform.

Pros

- Ensures high resource utilization at the infrastructure level.
- Reduces the capital investment on the infrastructure.
- Capital investment can be reduced.

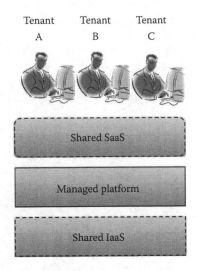

FIGURE 9.2
SaaS delivery from shared IaaS and managed platform.

Cons

- Still there is additional overhead in maintaining the development platform.
- Has to enable highly scalable and available features manually.
- There is a possibility of vendor lock-in at the infrastructure layer.

9.2.3 SaaS from Managed Infrastructure and PaaS

SaaS can be developed and delivered from self-managed infrastructure and shared PaaS. Normally, the application developed using this model will be deployed as an on-premise application. PaaS used here is generally a private PaaS. There are many PaaS providers who allow the customers to build their own private PaaS on their managed data center. This model will best suit the community deployment of SaaS. In the community deployment model, the infrastructure will be maintained by a group of organizations. The development platform (PaaS) can be accessed as a service by the different organizations. This type of model will reduce the overhead in maintaining the platform. SaaS-specific features such as high scalability and availability will be handled by the PaaS itself. But the overhead in maintaining the infrastructure will remain unsolved. This type of model also provides high security to the data. The vendor lock-in at the PaaS level can be possible in this type of model as the provider will be the third-party provider. This type of problem can be avoided by building our own PaaS platform on the managed

FIGURE 9.3
SaaS delivery from managed infrastructure and shared PaaS.

infrastructure. Figure 9.3 depicts the idea of developing and delivering SaaS from managed infrastructure and shared PaaS.

Pros

- The scalability and availability of the application will be provided by PaaS by default. So, the SaaS provider can concentrate more on application development.
- Security will be moderated, and there is full governance over user data.

Cons

- Even though overhead in maintaining the development platform is reduced, the overhead in maintaining the infrastructure still remains unsolved.
- This type of model will be suitable only for private/public SaaS applications. As the load on the public SaaS is high and unpredictable, the service providers may have to buy a new additional infrastructure to handle extra load. This is not possible for small SaaS development companies.

9.2.4 SaaS from IaaS and PaaS

This type of SaaS development and delivery model gets all the benefits of cloud computing. The infrastructure for developing and deploying a SaaS application will be provided by the IaaS provider, reducing the overhead in maintaining the underlying infrastructure. In the same way, the development platform also can be provided by the PaaS provider. The IaaS and PaaS provider might be public or private. The public IaaS and PaaS give more benefits than the private IaaS and PaaS. Normally, the public IaaS and PaaS will be selected to reduce the maintenance overhead and initial investment. Multitenancy is provided at all layers, that is, infrastructure, platform, and application layers. This type of multitenancy is called as high-level multitenancy, which is not available in other SaaS development and delivery models. Figure 9.4 illustrates the idea of developing and delivering SaaS from the IaaS and PaaS.

Pros

- The best delivery model that suits public SaaS applications.
- Ensures high resource utilization as it enables multitenancy at all layers of the application. It also supports Green IT applications.
- Dynamic scaling of IaaS and PaaS provider ensures the high scalability of the application. SaaS development companies need not worry about the scalability of the application.
- The high availability of the applications is ensured by the replica and the backup and recovery mechanism provided by the service provider.
- No overhead in maintaining the infrastructure and development platform. This enables SaaS development companies to develop more applications in a short span of time.

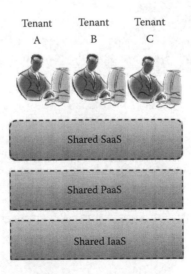

FIGURE 9.4
SaaS delivery from shared IaaS and PaaS.

Cons

- Since this type of model mostly uses public IaaS and PaaS models, the application will be hosted as off-premise applications. So, there is no governance over customer data.
- There is a possibility of cross tenant attacks as multitenancy is enabled at all levels of the application.

There are different development and deployment models discussed in this subsection. The deployment model of SaaS (public or private or community) will be selected based on the security requirement of the user data. The investment of buying and managing the infrastructure also will be considered before selecting the deployment model.

9.3 New Challenges

Many on-premise web applications are replaced by SaaS applications because of the business benefits of SaaS applications. Adoption of SaaS by large enterprises is increasing so as the adoption by individual users. This is due to the security issues of the data that are stored in the cloud. Large enterprises are worrying about data security, data governance, and availability. In this subsection, we shall discuss the challenges that make SaaS development difficult.

9.3.1 Multitenancy

Enabling multitenancy in a SaaS application is a challenge that all developers are facing today. Multitenancy can be achieved at the infrastructure, platform, database, and application levels for better resource utilization. Multitenancy is a one-to-many model that allows multiple customers to share a single instance of code and database of the SaaS application. So, developers need to learn the required knowledge of developing a multitenant software. For example, there are several multitenancy levels available with respect to database, namely, separate database, shared database and separate schema, and shared database and shared schema. The developer has to choose the correct multitenancy level based on the customer's requirement before developing the application. The developers can make use of any PaaS to reduce the overhead in enabling the multitenancy of the SaaS application.

9.3.2 Security

The first important consequence of the multitenancy model of the SaaS application is data security. Since the environment is shared, there is a possibility of data breaches. One tenant can easily get access to the other tenants' data, which will result in a serious issue. This is the reason why most of the enterprises are not ready to adopt SaaS applications for their business needs and isolating tenants' data is one way to overcome this, which is the biggest challenge for any SaaS developer. As SaaS applications are web based, the security threats that are applicable to traditional applications are applicable to SaaS applications as well. Internal and external security attacks should be detected and prevented by using proper intrusion detection and intrusion prevention systems. The developers can incorporate other security mechanisms such as strong encryption, auditing, authentication, access control, and authorization to secure SaaS applications.

9.3.3 Scalability

A scalable SaaS application should handle the extra load on the application efficiently. A SaaS application is said to be highly scalable if it handles the additional load properly. The scalability of the application can be maintained by predicting the load on the system and providing enough resources to handle the load. Since SaaS application users are diverse and can connect and disconnect from the system at any time, predicting the load on the application is a big challenge to the developer. So, the software architect should design an architecture that can handle any kind of load on the application. The architect should use vertical scaling, horizontal scaling, and load balancers to develop a highly scalable application. The scalability at the platform and infrastructure level also can decide the efficiency of the application. So, the application architecture should use IaaS and PaaS to utilize the advantages of cloud services.

9.3.4 Availability

Since SaaS users are storing their data in the service provider data center, ensuring the 99.99% availability of the data is a challenge to SaaS developers, and a better way to address this is to maintain a proper backup or replica mechanism. We cannot predict the failure of data stored in advance and cannot take preventive measures. Most of the traditional application developers rely on third-party tools to ensure the backup and disaster recovery. A developer can use distributed NoSQL databases that support automatic replication and disaster recovery rather than relational databases.

9.3.5 Usability

A new challenge introduced by the SaaS application is multiple-device support. The user may access the application from laptops, mobiles, tablets, and other handheld devices. Traditional web application developers used to develop rich Internet applications that are not adaptable to mobile devices.

So, by keeping the diversity of the end user devices, developers should develop a responsive web UI that adapts automatically to all user devices. The developers cannot develop a separate version of the application for different devices. The other challenge to the developer is to maintain the per tenant customization settings. The UI customization settings of each user should be managed properly for each device and should not affect the others.

9.3.6 Self-Service Sign-Up

Many traditional social networking sites use the feature of self-service sign-up for the users. But it is not mandatory for all the traditional web applications. But for the SaaS application, it is a mandatory feature to be incorporated, creating a challenge to the developers. SaaS applications should be available to the users as soon as they register. The application should not encourage any admin approval to allow the user to access the service. In the same way, the application should allow the users to unsubscribe from using the services at any point in time.

9.3.7 Automated Billing

The other challenge that is imposed by the unique characteristics of SaaS is automated billing. Most of the existing SaaS application users are facing the problem of abusive billing even after disconnected from the services. So, the developers should incorporate a proper mechanism to avoid abusive billing and maintain the usage history of each tenant. The tenant may subscribe to different services of the same service provider, and a mechanism should be available to the customers to see their total and per service usage report and billing.

9.3.8 Nondisruptive Updates

The frequent updates of the application may result in the unavailability of the application. Whenever the update is performed, it should not affect the normal behavior of the user. The other consequence of the update may lead to service-level agreement (SLA) violation. The update operation decides the downtime of the application, which is an important SLA parameter. The update should be performed in such a way that it does not increase the downtime that is mentioned in the SLA. So, scheduling and performing the nondisruptive updates of SaaS applications are the biggest challenges that developers face to avoid SLA violation.

9.3.9 Service Integration

A good SaaS application should be able to integrate with other third-party services. Normally, we cannot integrate third-party services directly. We have to use the APIs of service providers to perform service integration. So, the architecture of the SaaS application should allow service integration of other services through their APIs. Service integration can be achieved by following proper service-oriented architecture (SOA). Another benefit of service integration is that we can automate the implementation of some functionalities. Many SaaS providers fail to support service integration, which they need to improve while developing SaaS applications.

9.3.10 Vendor Lock-In

The major problem of public cloud services is vendor lock-in. There is no global standard followed among the service providers. Each service provider follows their own way of providing infrastructure and platform services, and the migration of the SaaS application from one service provider to the other becomes difficult and leads to vendor lock-in. So, developers should select the PaaS or IaaS that is interoperable with other service providers.

9.4 Cloud-Aware Software Development Using PaaS Technology

PaaS is a widely used cloud service model that enables the developers to develop an application online. PaaS provides the development PaaS on a 0 demand basis. The developers need not install any heavyweight software in their machine to use PaaS. The developers can develop and deploy an application online through the client tools such as web UI, Command Line Interface, web CLI, and representational state transfer Representational State Transfer (REST) API provided by the service provider. Normally, the development of SaaS application imposes a lot of challenges as discussed in the previous section. With the traditional development environment, it is very difficult to develop a successful SaaS that satisfies all the SaaS-specific requirements. To overcome these challenges, SaaS development companies can use the popular PaaS that provides many SaaS-specific features by default. By using PaaS, the developers can concentrate more on application functionalities rather than struggle with enabling SaaS-specific features. In this subsection, we shall discuss developing cloud-aware SaaS applications using PaaS technology.

Benefits of PaaS: PaaS is used by many small SaaS development companies and ISVs. The following discusses the many characteristics that increase PaaS adoption by an organization:

- PaaS provides the services to develop, test, deploy, host, and maintain applications in one place.
- Most of the service providers offer *polyglot* PaaS where the developers can use a variety of application development environments in one integrated development environment (IDE).

- The variety of client tools such as web UI, web CLI, REST APIs, and IDE integration increases the ease of application development and deployment.
- PaaS offers a built-in multitenant architecture for the applications developed using PaaS.
- PaaS provides a software load balancer that ensures the dynamic scaling of the application.
- The replicas maintained by PaaS providers ensure the high availability of the application.
- PaaS providers also allow integrating with other web or cloud services to develop composite cloud services.
- PaaS increases the collaboration between the development team as the application will be deployed at a central place.
- The other important SaaS-specific features like monitoring tools and automated billing will be offered by PaaS itself.

Before PaaS and after PaaS: PaaS changes the software development process totally when compared to the traditional software development model. In the traditional software development, the development process will start from requirements analysis that involves all the stakeholders of the system. The second phase is the design phase that includes software architecture, UI, and database design. The implementation phase is the actual development, or coding the application with the available development platform. Here, the development platform will be a license-based heavyweight software that requires machines with high computing power, forcing companies to invest more on the development platform and hardware. The testing is the next phase of software development that mainly ensures the nonfunctional requirements like security, scalability, availability, and portability. There are many tools available in carrying out the testing process of the developed application. After testing a product, it can be deployed in suitable infrastructure and can be delivered to the end users. Normally, if the application is a stand-alone application, it will provide a license. If it is a web application, it will be delivered through the Internet. Generally, each customer of the application will get a separate copy of the application. In the case of a web application, the application will be hosted in the service provider infrastructure or customer on-premise infrastructure. The next step is to maintain the scalability and availability of the application. The company should keep enough additional servers to handle the extra load of the application. But in real time, the load on the application will be dynamic and unpredictable. So, we cannot decide the power of the hardware, which we need to add later. Here, buying and maintaining the additional hardware will be an overhead to development companies. The next important thing is availability of the application. To ensure high availability, companies need to keep replicas. Again replica management is a big issue for the companies. In the end, the maintenance process of the application will be carried out by companies. Normally, updates will be performed by customers from their machines. Each copy of the application should be updated separately. Additionally, sometimes updates through a slow Internet connection will disrupt the normal behavior of the system. So, to avoid all these problems in traditional software development methodologies, companies started using the PaaS technology for better productivity.

Figure 9.5 illustrates the software development process before and after PaaS. The main problem in traditional software development is the licensed platform, overhead in testing, deployment, scaling, and maintenance of the application. But PaaS hides all of the overheads that are present in traditional software development. Here, the PaaS platform will be provided to the developers on an on-demand basis through the Internet. So, the developers can use the platform to develop their application. The developed application will be automatically deployed on the service provider infrastructure by the PaaS tool itself. The other important parameters such as dynamic scaling and availability will be handled by the PaaS tool. Sometimes, because of poor Internet connection, the developers may not able to use PaaS online. So, to overcome this problem, some of the PaaS vendors allow the offline software development by

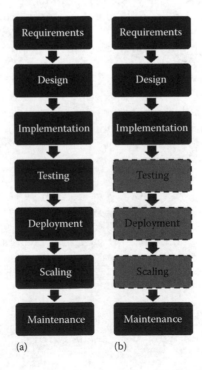

(a)   (b)

FIGURE 9.5
Software development (a) before and (b) after PaaS.

integrating their online repository with the local IDEs. This enables the developers to work offline in the local machine and push the application online whenever there is an Internet connection. The other advantage is centralized maintenance. Here, the application is hosted in a central location. So, the customers need not perform the updates from their machine as the service provider updates the application that will be effected to all the customers.

The developed SaaS application is different from the traditional application. SaaS application developers should enable the following features to the application: multitenancy, dynamic scaling, and high availability. The process of developing a SaaS application using PaaS technology is discussed in the following.

9.4.1 Requirements Analysis

The development team should cope up with frequently changing requirements of the SaaS application. Generally, in requirements analysis, only the customers and the development team will be involved. But in SaaS application development, the service provider for the PaaS tool should also be involved. The requirements collection team should collect the requirements from all the stakeholders. The requirement document should include functional, nonfunctional, and other SaaS-specific requirements. Before developing the application, the requirements collection team should analyze the suitability of the SaaS delivery model for the customers' requirements. Generally, the SaaS delivery model will be selected based on security and ROI. The next important thing in the requirements analysis phase is the deployment model of the SaaS application. If the application does not need more security, we can develop the SaaS application from any public PaaS provider. If it requires more security, then we have to select any private PaaS provider to develop the application. Normally, in the public deployment model, the overhead in maintenance will be low and security threats will be high. But in the private deployment model, overhead in maintenance will be high and security threats will be moderate. As the SaaS application is going to be delivered through the Internet, the SaaS development company should assign highly skilled security experts and software architects in order to succeed in the SaaS development business. Security experts should ensure the security at all layers of the application. The software architect should ensure the scalability and availability of the application. Once the requirements analysis is properly done, the development team may start designing the architecture.

9.4.2 Multitenant Architecture

An important characteristic of the SaaS application is multitenancy where multiple customers are allowed to share the same application. Achieving multitenancy depends on the software architecture. The software architecture should ensure the multitenancy of the SaaS application. Multitenancy can be achieved at the infrastructure, development platform, database, and application

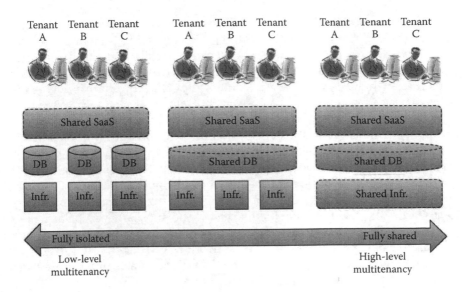

FIGURE 9.6
Different multitenancy levels.

levels. The architect can choose different levels of multitenancy to utilize the resources effectively. If the software architect selects an IaaS provider for the infrastructure needs, then the architect is relieved from enabling multitenancy at the infrastructure level. If the architect chooses the PaaS provider, the platform level and the infrastructure level multitenancy will be provided by the PaaS provider itself. So, the software architect's job gets reduced to enable the multitenancy features only at the software level. Depending on the multitenancy level, the isolation security of the data is decided. If multitenancy is achieved at all levels, then it is called as high-level multitenancy. If multitenancy is achieved only at the application level, then it is called as low-level multitenancy. The security threat to data will increase as the multitenancy level increases. Depending on the user security requirements, the architect should select the multitenancy level. The other advantage of multitenancy is resource utilization. If the company wants high resource utilization, they can go for a high-level multitenancy. Figure 9.6 illustrates the different levels of multitenancy and isolation available to ensure resource utilization and security to the software architect.

9.4.3 Highly Scalable and Available Architecture

Another important characteristic of the SaaS application is dynamic scaling. The dynamic scaling feature is not mandatory in the case of traditional web applications. But in the case of the SaaS application, dynamic scaling is very important.

Like multitenancy, achieving dynamic scaling also depends on the software architecture. As the load on the SaaS application becomes unpredictable and increases or decreases any time, the architecture should ensure the same performance on varying loads. The scalability of the SaaS application can be achieved using horizontal scaling, vertical scaling, software load balancer, and hardware load balancer. In horizontal scaling, identical resources (application server, database server, and infrastructure) will be added to the application to handle the additional load. In vertical scaling, the capacity of the server (application, database, and infrastructure) will be increased as the load increases. The software load balancers also can be used to ensure the dynamic scalability of the SaaS application. The role of the software load balancer is to distribute additional user request across different application and database servers. The hardware load balancer will distribute the load across different virtual machines when there is a need for more computing power. If the infrastructure and development platform is consumed from any service providers (IaaS, PaaS), they will provide the hardware and software load balancers to balance the load. If the platform and the infrastructure are self-managed, then the SaaS development company should rely on third-party tools or they have to develop their own. Figure 9.7 illustrates the typical SaaS architecture used to achieve high scalability and availability of the SaaS application.

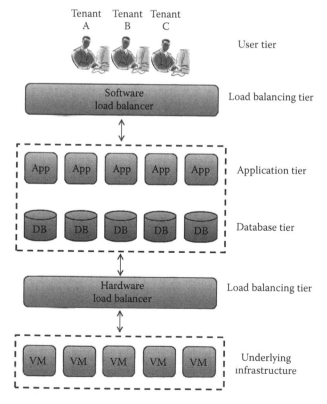

FIGURE 9.7
Typical architecture of the SaaS application.

Like multitenancy and scalability, the availability of the application is also an important characteristic of a SaaS application. The availability of the application decides its uptime and downtime. The availability of the application is an important parameter of the SLA. Ensuring the 99.99% availability of the application depends on the replica mechanism that is specified in the software architecture. As shown in Figure 9.7, multiple copies of virtual machines and database applications should be maintained for high availability. While maintaining the replica, another important aspect is recovery time after any failure. The application should be fault tolerant, and the recovery time should be minimal to avoid SLA violation. The replica should be maintained near the customer location to reduce the recovery time after any failure or disaster.

9.4.4 Database Design

Achieving multitenancy, scalability, and availability at the database level is an important criterion for successful SaaS development. The database design for multitenancy should consider the security requirement of the data.

Multitenancy at database level can be achieved by sharing the database instance, sharing the database table, and sharing the database schema. Depending on the security of the application, the database should be designed to secure multitenancy. The database-level multitenancy can be achieved in three different ways as illustrated in Figure 9.8. If the database designer selects a separate database for different tenants as shown in Figure 9.8b, the security will be ensured. If the shared database and separate schema are

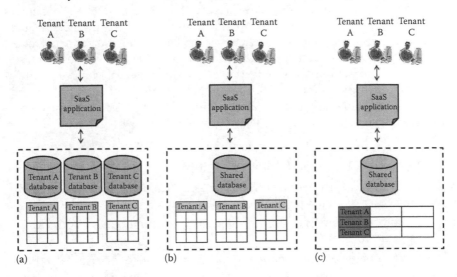

FIGURE 9.8
Database-level multitenancy: (a) separate database, (b) shared database and separate schema, and (c) shared database and shared schema.

selected as shown in Figure 9.8a, the security to the data will be moderated. The third multitenancy model shares the database and the schema for the tenants as shown in Figure 9.8c.

The scalability and availability of the database decide the performance of the application. Most of the SaaS applications are interactive and involve a large number of database of read and write requests from the users. When the number of requests exceeds the actual capacity of the database server, additional requests should be redirected to the other server. When requests are redirected, a change of data in one database server should reflect in other database servers also. Mostly the type of data used in the SaaS application will be diverse and includes structured, semistructured, and unstructured data in huge amounts. So, for SaaS applications, Not Only Structured Query Language (NoSQL) databases will be a better option than traditional relational, object-oriented databases. The developers can leverage the advantages of NoSQL databases to achieve scalability and availability at the database level in an efficient way.

9.4.5 SaaS Development

After designing the architecture and the databases, the developers need to implement the functional requirements given by the customers. As we discussed earlier, PaaS facilitates the developers in developing highly scalable, available, and multitenant-aware SaaS applications.

The PaaS tools allow the developers to develop the application online, and the application will be deployed on the service provider infrastructure as soon as the developer pushes the application online. Here, the end users or SaaS consumers can access the application online using the web UI provided by the SaaS provider. Figure 9.9 illustrates the overview of SaaS development using PaaS tools.

The PaaS providers also provide testing tools in the same development environment to facilitate the developers. So, the developers can use built-in testing tools provided by the PaaS providers to test SaaS applications. Some PaaS providers offer automated testing of the applications also. While developing the application, the developers should incorporate the following things for successful SaaS:

- Responsive UI design to support multiple devices
- Role Based Access Control (RBAC), Access Control List (ACL) mechanism to uniquely identify users and tenants
- Monitoring tools that will monitor the performance and notify the service provider frequently
- Control panel for the tenant and admin to manage the users
- User-centric customization panel that does not affect the settings of other tenants or users
- Self-service sign-up for the users

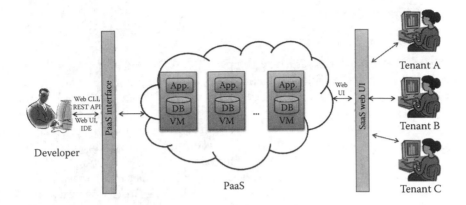

FIGURE 9.9
Overview of SaaS development using PaaS.

- Usage statistics and bill calculation
- Help documentation to use the service
- Service integration

9.4.6 Monitoring and SLA Maintenance

As soon as the SaaS application is developed and deployed using PaaS, the end users can access the SaaS application over the Internet from any end user device such as desktops, laptops, tablets, and mobiles. After delivering the application, any misbehavior, failure, security attacks, and disasters of SaaS applications should be monitored and prevented. Since a lot of customers are sharing the same instance of a single application, any misbehavior from one tenant will affect the other tenants and the underlying resources. Updates should be scheduled and performed in such a way that it does not affect the normal behavior of the system, so updating the SaaS application frequently will offer the its most updated version to the end users. But if you update the application with bulk updates frequently, it may lead to the unavailability of the application. Another important job in SaaS monitoring is to monitor the SLA violation by both the service provider and customer. If there is any SLA violation, the monitoring tool should notify the developers to correct the errors that lead to the SLA violation. The SaaS providers should define the SLA clearly to the end users before delivering any services. The SLA should include the availability, response time, and degree of support. The service providers also should provide 24 × 7 support to the end users. The development team should resolve the issues frequently as soon as feedback is received from the end users. There are many third-party monitoring tools are available to monitor SaaS applications. The SaaS development company can make use of those monitoring tools to reduce the overhead.

9.5 Summary

SaaS is one of the important services provided by cloud computing. Usage-based billing, high scalability, ease of access, and other benefits make the most of the customers to move from traditional applications to SaaS applications. Small business companies/start-up companies started using SaaS to reduce their investment on buying software that is underutilized in their organization. By using on-demand SaaS applications, any company can increase their ROI. If you look at the usage of SaaS in large enterprises, it is very low compared to the usage of individuals and small business companies. Large enterprises are hesitant to use SaaS applications for their organizations because of security issues. SaaS applications are multitenant. Whenever the users are sharing the application, there is a possibility of security attacks between the tenants. If you remove the multitenant features from SaaS applications at the infrastructure, platform, and software levels, it will result in a high development cost. Obviously, the customers need to pay more for the software they use. When someone decides to go for SaaS applications, they have to consider its cost and security requirements. Based on the cost and security requirements, the service providers can follow any of the SaaS development and deployment models discussed in this chapter. Other than the security issues, the SaaS application introduces a lot of challenges to the developers such as scalability, availability, usability, self-service sign-up, and automated billing. These challenges can be addressed by incorporating the best practices into software engineering and PaaS technology. SaaS changes the way the software is delivered, and PaaS changes the way the software is developed. PaaS automates the process of deployment, testing, and scaling and reduces manual work and the cost involved in developing the application. SaaS providers also can utilize IaaS of cloud computing to reduce the investment on buying infrastructure.

Review Points

- *SaaS* is one of the software delivery models that allow the end users to share the application that is centrally hosted by a provider (see Section 9.1).
- *SaaS* contains unique characteristics that differentiate it from traditional software (see Section 9.1.1).
- *SaaS benefits* include pay per use, zero infrastructure, ease of access, automated updates, and composite services (see Section 9.1.2).
- *SaaS does not suit* the application where fast processing of data is needed (see Section 9.1.3).

- *SaaS delivery* can be of many forms: managed infrastructure and platform, IaaS and managed platform, managed infrastructure and PaaS, and IaaS and PaaS (see Section 9.2).
- *SaaS challenges* such as multitenancy, security, scalability, availability, and usability make SaaS development a difficult job for developers (see Section 9.3).
- *Multitenancy* is a one-to-many model where a single instance of an application can be shared by multiple users (see Section 9.3.1).
- *Scalability* of the SaaS application depends on how well the application will handle the extra load (see Section 9.3.3).
- *Availability* of the SaaS application can be improved by keeping proper backup and recovery mechanisms (see Section 9.3.4).
- *Usability* of the SaaS application depends on the adaptive and responsive UI design that supports multiple devices (see Section 9.3.5).
- *Self-service sign-up* feature of the SaaS application allows the end users to subscribe or unsubscribe from the service without the intervention of the provider (see Section 9.3.6).
- *Automated billing* feature maintains the usage history and provides the bill based on per tenant usage or per service usage (see Section 9.3.7).
- *Nondisruptive updates* ensure the uptime of the application and during the time of application update also (see Section 9.3.8).
- *Service integration* of the SaaS application allows any SaaS application to integrate with other services through an API (see Section 9.3.9).
- *Vendor lock-in* does not allow migration of application to other service providers, which is the problem with most of the public cloud providers (see Section 9.3.10).
- *PaaS* changes the way software is developed by providing development PaaS (see Section 9.4).
- *Cloud-aware software development* requires multitenant, highly scalable architecture (see Section 9.4).

Review Questions

1. What is Software as a Service (SaaS)? How is it different from traditional software?
2. Briefly explain the benefits of the SaaS application.
3. Is it wise to choose the SaaS delivery model for all kinds of applications? Justify your answer.

4. Explain the different SaaS development and deployment models with neat diagrams.
5. List out the pros and cons of different SaaS development and deployment models.
6. List the challenges that make SaaS development a difficult task. Also, explain any five challenges in detail.
7. Write short notes on the benefits provided by PaaS technology for developing SaaS applications.
8. Explain in detail how PaaS technology changes software development.
9. Briefly explain the requirements analysis for SaaS application.
10. Explain different multitenancy levels with neat diagrams.
11. Illustrate and explain the typical architecture of the SaaS application, which is to ensure better scalability and high availability.
12. How is database-level multitenancy achieved? Explain the different database-level multitenancies with neat diagrams.
13. List out important features that SaaS developers should incorporate while developing SaaS applications.
14. Write short notes on monitoring and SLA maintenance of SaaS applications.

Further Reading

6 best practices to cloud enable your apps. White Paper, Tier 3, Inc.
Best practices for cloud computing multi-tenancy. White Paper, IBM Corporation, 2003.
Betts, D., A. Homer, A. Jezierski, M. Narumoto, and H. Zhang. *Developing Multi-Tenant Applications for the Cloud on the Microsoft Windows Azure*. Microsoft Press, 2010.
Building successful enterprise SaaS apps for the cloud. White Paper. THINKstrategies, Inc., 2011.
Chauhan, N. S. and A. Saxena. A green software development life cycle for cloud computing. *IT Professional* 15(1): 28–34, 2013.
Chong, R. F. Designing a database for multi-tenancy on the cloud: Considerations for SaaS vendors. Technical article, IBM Developer Works. Available [Online]: http://www.ibm.com/developerworks/data/library/techarticle/dm-1201dbdesigncloud/dm-1201dbdesigncloud-pdf.pdf Accessed October 12, 2013.
da Silva, E.A.N. and D. Lucredio. Software engineering for the cloud: A research roadmap. *26th Brazilian Symposium on Software Engineering (SBES)*, September 23–28, 2012, pp. 71–80.
Deploying software as a service (SaaS). White Paper, WebApps, Inc. a.k.a. SaaS.com.

Gagnon, S., V. Nabelsi, K. Passerini, and K. Cakici. The next web apps architecture: Challenges for SaaS vendors. *IT Professional* 13(5): 44–50, 2011.

Goth, G. Software-as-a-service: The spark that will change software engineering? *Distributed Systems Online, IEEE* 9(7): 3, 2008.

Kang, S., J. Myung, J. Yeon, S. Ha, T. Cho, J. Chung, and S. Lee. A general maturity model and reference architecture for SaaS. *Proceedings of Database Systems for Advanced Applications (DASFAA 2010)*, Part II, LNCS 5982, pp. 337–346.

Lawton, G. Developing software online with platform-as-a-service technology. *Computer* 41(6): 13–15, June 2008.

Liu, F., J. Tong, J. Mao, R. B. Bohn, J. V. Messina, M. L. Badger, and D. M. Leaf. NIST cloud computing reference architecture. NIST Special Publication 500-292, September 2011. Available [Online]: http://www.nist.gov/customcf/get_pdf.cfm?pub_id=909505 (accessed September 3, 2013).

Mell, P. and T. Grance. The NIST definition of cloud computing. NIST Special Publication 800-145, September 2011. Available [Online]: http://csrc.nist.gov/publications/nistpubs/800-145/SP800-145.pdf (accessed September 3, 2013).

Yau, S. S. and H. G. An. Software engineering meets services and cloud computing. *Computer* 44(10): 47–53, October 2011.

Rodero-Merino, L., L. M. Vaquero, E. Caron, A. Muresan, and F. Desprez. Building safe PaaS clouds: A survey on security in multitenant software platforms. *Computers and Security* 31(1), 96–108, 2012. ISSN 0167-4048.

SaaS Architecture. White Paper, Progress Software Corporation.

SaaS Scalability. White Paper, Progress Software Corporation.

10

Networking for Cloud Computing

Learning Objectives

After studying this chapter, you should be able to

- Understand the general classification of data centers
- Present an overview of the data center environment
- Understand the basic networking issues in data centers
- Explain the performance challenges faced by TCP/IP in data center networks
- Describe the newly designed TCPs for data center networks and their novelty

Preamble

This chapter provides an introduction to networking in Cloud Enabled Data Centers (CEDCs) and the issues thereof. A general classification of data centers and a brief overview of the data center environment are provided to familiarize the reader with the CEDCs. Major issues related to networking in a cloud environment are presented with an emphasis on TCP/IP-related performance issues. Newly designed protocols tailored specifically for data center networks are explained in detail, while mentioning advantages and disadvantages of each.

10.1 Introduction

The Internet over the past few years has transformed from an experimental system into a gigantic and decentralized source of information. Data centers form the backbone of the Internet and host diverse applications ranging from social networking to web search and web hosting to advertisements. Data centers are mainly classified into two types [1]: the ones that aim to provide online services to users, for example, Google, Facebook, and Yahoo, and others that aim to provide resources to users, for example, Amazon Elastic Compute Cloud (EC2) and Microsoft Azure.

Data centers in the recent past have transformed computing, with large-scale consolidation of enterprise IT into data center hubs and with the emergence of several cloud computing service providers. With the widespread acceptance of cloud computing, data centers have become a necessity. Since cloud computing is becoming an important part of the foreseeable future, studying and optimizing the performance of data centers have become extremely important.

Building an efficient data center is necessary in order to strengthen the data processing and to centrally manage the IT infrastructure. However, ever since the inception of data centers, their operation and maintenance have always been a complex task. It is only after 1994 that the usage of data centers increased extensively. Based on the fault-tolerance capacity and service uptime, today's data centers are classified into four tiers as shown in Table 10.1.

Tier I–IV is a standard methodology used to define the uptime of a data center. This is useful for measuring data center performance, investment, and return on investment (ROI). Tier IV data center is considered to be most robust and less prone to failures. It is designed to host mission critical servers and computer systems with fully redundant subsystems (cooling, power, network links, storage, etc.) and compartmentalized security zones controlled by biometric access control methods. The simplest is Tier I data center, which is usually used by small shops.

As maintaining data centers involves a lot of complexity and cost, small- and medium-range companies cannot build their own data centers. Thus, companies like Google, Amazon, Microsoft, Facebook, and Yahoo transformed into cloud computing service providers and started building Internet data centers (IDCs) to meet the scaling demands of the cloud users.

Today, the data centers of the aforementioned companies host diverse applications such as web search, web hosting, social networking, storage, e-commerce, and large-scale computations. As the variety, complexity, and

TABLE 10.1

Classification of Data Centers

Tiers	Features	Uptime (%)
I	Nonredundant capacity components (single uplinks and servers)	99.671
II	Tier I + redundant capacity components	99.741
III	Tier I + Tier II + dual-powered equipments and multiple links	99.982
IV	Tier I + Tier II + Tier III + all components are fault tolerant including uplinks, storage, HVAC systems, servers + everything is dual powered	99.995

TABLE 10.2

Guide to Where Costs Go in a Data Center

Amortized (%)	Cost Component	Subcomponents
45	Servers	CPU, memory, storage systems
25	Infrastructure	Power distribution and cooling
15	Power draw	Electricity utility costs
15	Network	Links, transit, equipment

Source: Greenberg, A. et al., *SIGCOMM Comput. Commun. Rev.*, 39(1), 68, December 2008.

penetration of such services grow, data centers continue to expand and proliferate. Majority of the data centers, however, face the daunting challenges of reducing the server cost, infrastructure cost, and excessive power consumption and optimizing the network performance. Table 10.2 [2] shows where the costs go in today's cloud service data centers.

This chapter highlights the challenges faced toward designing fast and efficient networks for communication within cloud-enabled data centers (CEDCs). The major emphasis, however, is toward understanding the transport layer issues in data center networks (DCNs).

10.2 Overview of Data Center Environment

Initially, the organizations used to maintain *server rooms*. Generally, server rooms used to house servers and the necessary network electronics for establishing a LAN. Some organizations used to provision one main and one standby server room. The standby server room was equipped to maintain the necessary functions in the event of the main room being put out of action. For better fault tolerance, a few organizations opted to locate these rooms in different buildings.

The advent of client–server computing and the Internet led to the concept of data centers. Large data center companies thrived in the dot-com era when Internet companies faced rapid growth. The data centers were focused on providing reliability. Since then, data center paradigm has served as the foundation for information technology that either runs business or is the business. That paradigm has been evolutional throughout the last several decades and transformational in the past 5–10 years.

The description of a data center has almost always been preceded with *mission critical*, because that is the service it provides—the mission critical hardware and software where maximum uptime is required. The data center is a fortress, dedicated to achieving maximum reliability at any cost. While reliability is still the key factor, the data center has evolved and advancements in the past 5 years have accelerated the pace of innovation.

10.2.1 Architecture of Classical Data Centers

The data center is home to the computational power, storage, and applications necessary to support an enterprise business. The data center infrastructure is central to the IT architecture, from which all content is sourced or passes through. Proper planning of the data center infrastructure design is critical, and performance, resiliency, and scalability need to be carefully considered.

The *multitier* model is the most common design in the enterprise. It is based on the web, application, and database layered design supporting commerce and enterprise business ERP and CRM solutions. This type of design supports many web service architectures, such as those based on Microsoft. NET or Java 2 Enterprise Edition. These web service application environments are used by ERP and CRM solutions from Siebel and Oracle, to name a few. The multitier model relies on security and application optimization services to be provided in the network.

The *server cluster* model has grown out of the university and scientific community to emerge across enterprise business verticals including financial, manufacturing, and entertainment. The server cluster model is most commonly associated with high-performance computing (HPC), parallel computing, and high-throughput computing (HTC) environments but can also be associated with grid/utility computing. These designs are typically based on customized, and sometimes proprietary, application architectures that are built to serve particular business objectives.

Multitier model: The multitier data center model is dominated by HTTP-based applications in a multitier approach. The multitier approach includes web, application, and database tiers of servers. Today, most web-based applications are built as multitier applications. The multitier model uses software that runs as separate processes on the same machine using interprocess communication (IPC), or on different machines with communications over the network. Typically, the following three tiers are used:

1. Web server
2. Application
3. Database

Multitier server farms built with processes running on separate machines can provide improved resiliency and security. Resiliency is improved because a server can be taken out of service while the same function is still provided by another server belonging to the same application tier. Security is improved because an attacker can compromise a web server without gaining access to the application or database servers. Web and application servers can coexist on a common physical server; the database typically remains separate.

10.2.2 CEDCs

New technological and economic pressures have forced organizations to look at ways to get more out of their IT infrastructure. The current state of IT infrastructure is strained, and the new demands make it all the more difficult for businesses to maintain efficiency and effectiveness. This has a bearing on the quality of services irrespective of the number of users and applications. The solution lies in cloud computing. The cloud has the capability to reduce costs and increase the flexibility of applications and services including IT infrastructure. The CEDC takes a virtualized data center (and business) into one that is more agile because it takes virtualization to the next level. The virtualized environment is transformed into one that is optimized with intelligent, integrated IaaS, and PaaS that manages dynamic workloads—giving the workloads the resources that it needs based on business policies. It also provides automation and orchestration of resources across heterogeneous data centers. This progression from a virtualization management to a CEDC is important as it addresses common business goals we hear over and over again.

10.2.3 Physical Organization

Data centers generally span across the entire building, a few floors of a building or even a single room in the building. Figure 10.1, collected from Google images, shows one of the several possible ways of arranging the server racks in a data center. Data centers usually comprise of a large number of servers that are mounted in rack cabinets and are placed in single rows forming corridors (so-called aisles) between them, so as to allow access to the front and rear of each cabinet.

FIGURE 10.1
Physical organization of a data center.

Moreover, a few equipments such as storage devices are often as large as the racks. Such equipments are generally placed alongside the racks. Large data centers house several thousand servers wherein sometimes shipping containers packed with 1000 or more servers each are used. In the event of a failure or when upgrades are required, the entire containers are replaced rather than replacing an individual server.

10.2.4 Storage and Networking Infrastructure

Typically, data centers require four different types of network accesses and, hence, could use four different types of physical networks as shown below:

1. *Client–server network*: To provide external connectivity to the data center. Traditional wired Ethernet or Wireless LAN technologies can be used.

2. *Server–server network*: To provide high-speed communication among the servers of the data center. Ethernet, InfiniBand (IBA), or other technologies can be used. Figure 10.2 shows an example of server–server network.

3. *Server–storage network*: To provide high-speed connectivity between the servers and storage devices. Usually Fiber Channel is used, but technologies like Ethernet or InfiniBand can also be used.

4. *Other networks* such as a network required to manage the data center. Generally, Ethernet is used but the cabling may be different from the mainstream networks.

Multiple applications run inside a single data center, typically with each application hosted on its own set of (potentially virtual) server machines.

FIGURE 10.2
Networking infrastructure in data centers.

Each application is associated with one or more publicly visible and routable IP addresses to which clients in the Internet send their requests and from which they receive replies. Inside the data center, requests are spread among a pool of frontend servers that process the requests. This spreading is typically performed by a specialized load balancer.

A two-tier network topology is very popular in DCNs today. Access switches for server connectivity are collapsed in high-density aggregation switches that provide the switching and routing functionalities for access switching interconnections and various LAN servers. It has several benefits:

- Design simplicity (fewer switches and so fewer managed nodes)
- Reduced network latency (by reducing the number of switch hops)
- Typically a reduced network design oversubscription ratio
- Lower aggregate power consumption

However, a disadvantage of a two-tier design includes limited scalability: when the ports on an aggregation switch pair are fully utilized, then the addition of another aggregation switch/router pair adds a high degree of complexity. The connection between aggregation switch pairs must be fully meshed with high bandwidth, so no bottlenecks are introduced into the network design. Since an aggregation switch pair is also running routing protocols, more switch pairs means more routing protocol peering and more routing interfaces and complexity introduced by a full mesh design.

10.2.5 Cooling Infrastructure

Since the data centers usually span across the entire building and house several thousand servers, a sophisticated cooling infrastructure is deployed, which may involve building level air-conditioning units, fans, and air recirculation systems. Figure 10.3 shows one of the possible arrangements of aisles by which the cooling infrastructure is simplified.

The server racks are placed on a raised plenum and arranged in alternately back-facing and front-facing aisles. Cold air is forced up in the front-facing aisles, and the server or chassis fans draw the cold air through the server to the back. The hot air on the back then rises and is directed (sometimes by using some deflectors) toward the chiller plant for cooling and recirculation. The author in [3] mentions that although such a setup is not expensive, it can create hot spots either due to uneven cooling or the mixing of hot and cold air.

In recent years, there have been many innovations in power and cooling technologies and management of facilities. Efficiencies have been integrated into every aspect of the data center and building design, covering everything

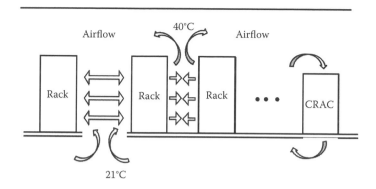

FIGURE 10.3
Cooling in a data center. (From Kant, K., *Comput. Netw.*, 53(17), 2939, 2009.)

from bunkers to chicken coop design and mobile data centers to using the building as an air handler. Green technologies and environmental awareness have also been a large part of the industry in the past 3 years. No longer just a choice between build and lease, the data center can be owned, placed in colocation, wholesale, put in a public or private cloud, or a hybrid strategy of options.

The changes that have come about have even altered the meaning of what constitutes a data center. To Google, Microsoft, or Yahoo, it is a hyperscale facility with tremendous innovation engineered into it. For the consolidation projects, it means taking what they once considered to be data centers and bringing them into a small number of new, large-scale facilities. To others, their definition of a data center was transformed by the advances in IT equipment that required more power, more cooling, and a more advanced facility to support it.

Many facets of site selection for data centers were shaped by the proliferation of fiber networks and the need to both avoid natural disasters and achieve extreme power and cooling efficiencies. In the United States, several regions became data center hubs, where Internet companies and enterprises are selected to build new data centers. Silicon Valley continued to prosper and grow, and regional hubs developed in Quincy, Washington, Chicago, Dallas/Fort Worth, North Carolina, and the New York/New Jersey region.

10.2.6 Nature of Traffic in Data Centers

Data center environment is largely different from that of the Internet, for example, the round-trip time (RTT) in DCNs can be as less as 250 μs in the absence of queuing [4]. The reason is that DCNs are privately owned networks tailored to achieve high bandwidth and low latency. Moreover,

TABLE 10.3

Data Center Traffic: Applications and Performance Requirements

Traffic Type	Examples	Requirements
Mice traffic (<100 kB)	Google search, Facebook	Short response times
Cat traffic (100 kB to 5 MB)	Picasa, YouTube, Facebook photos	Low latency
Elephant traffic (>5 MB)	Software updates, video on demand	High throughput

Source: Reproduced from Kant, K., *Comput. Netw.*, 53(17), 2939, December 2009.

the nature of traffic in DCNs largely varies from that of the Internet traffic. Traffic in DCNs is classified mainly into three types [5]:

1. *Mice traffic*: The queries form the mice traffic (e.g., Google search and Facebook updates). Majority of the traffic in a DCN is query traffic, and its data transmission volume is usually less.

2. *Cat traffic*: The control state and coordination messages form the cat traffic (e.g., small- and medium-sized file downloads)

3. *Elephant traffic*: The large updates form the elephant traffic (e.g., antivirus updates and movie downloads).

The different traffic types in DCNs, their applications, and performance requirements are summarized in Table 10.3. Thus, bursty query traffic, delay-sensitive cat traffic, and throughput-sensitive elephant traffic coexist in DCNs.

10.3 Networking Issues in Data Centers

Cloud networking is the fundamental backbone to provide cloud services and the reason behind the shift in how IT services are provided to users. This section focuses on several issues that are faced in cloud networking.

10.3.1 Availability

One of the daunting challenges that a cloud provider organization faces is to provide maximum uptime for the services that are offered to the users. Even a few seconds of downtime may lead to loss of reputation for the organization and affect the overall business. Moreover, downtime may lead to violation of service-level agreements (SLAs) between the cloud user and the cloud provider, thus largely affecting the cloud provider's revenues. The most widely adopted approach to achieve high availability is to replicate the data and take regular backups.

10.3.2 Poor Network Performance

The three basic performance requirements of a DCN are high burst tolerance, low latency, and high throughput [5]. This is because traffic in a data center comprises a mix of all the three kinds of traffic: mice traffic, cat traffic, and elephant traffic, each having a different performance requirement than the other.

The major challenge is that traditional Transmission Control Protocol/ Internet Protocol (TCP/IP) stack, which is mainly designed for Internet-like scenarios, fails to provide optimal performance in DCNs. The next section provides a more detailed description about several TCP/IP issues in DCNs.

10.3.3 Security

Keeping a cloud user's data secure during transit, or while it is at rest, is yet a concern for the cloud service providers. Accidental deletion of data because of a power outage or malfunctioning of a regular backup program may lead to loss of customer's data and incur huge damage for the hosting organization.

Apart from the aforementioned concerns, cloud providers also need to ensure physical security of a data center building and its networking infrastructure to prevent any attacks from malicious insiders.

10.4 Transport Layer Issues in DCNs

TCP is one of the most dominant transport protocols, widely used by a large variety of Internet applications, and also constitutes majority of the traffic in both types of DCNs [5]. It has been the workhorse of the Internet ever since its inception. The success of the Internet, in fact, can be partly attributed to the congestion control mechanisms implemented in TCP. Though the scale of the Internet and its usage increased exponentially in the recent past, TCP has evolved to keep up with the changing network conditions and has proven to be scalable and robust.

However, it has been observed that the state-of-the-art TCP fails to satisfy the three basic requirements (mentioned in the previous subsection) together within the time boundaries because of impairments such as TCP incast [4], TCP outcast [6], queue buildup [5], buffer pressure [5], and pseudocongestion effect [7], which are discussed further in the following sections.

10.4.1 TCP Impairments in DCNs

Although TCP constantly evolved over a period of three decades, the diversity in the characteristics of present and next-generation networks and a variety of application requirements have posed several challenges to TCP congestion control mechanisms. As a result, the shortcomings in the fundamental design of TCP have become increasingly apparent. In this section, we mainly focus on the challenges faced by the state-of-the-art TCP in DCNs.

10.4.1.1 TCP Incast

TCP incast has been defined as the pathological behavior of TCP that results in gross underutilization of the link capacity in various many-to-one communication patterns [8], for example, partition/aggregate application pattern as shown in Figure 10.4. Such patterns are the foundation of numerous large-scale applications like web search, MapReduce, social network content composition, and advertisement selection [5,9]. As a result, TCP incast problem widely exists in today's data center scenarios such as distributed storage systems, data-intensive scalable computing systems, and partition/aggregate workflows [1].

In many-to-one communication patterns, an aggregator issues data requests to multiple worker nodes. The worker nodes, upon receiving the request, concurrently transmit a large amount of data to the aggregator (see Figure 10.5). The data from all the worker nodes traverse a bottleneck link in many-to-one fashion. The probability that all the worker nodes send the reply at the same time is high because of the tight time bounds. Therefore, the packets from these nodes happen to overflow the buffers of top-of-the-rack (ToR) switches and, thus, lead to packet losses. This phenomenon is known as synchronized

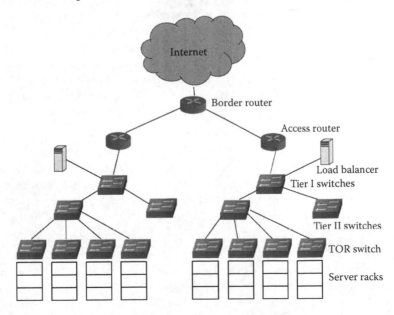

FIGURE 10.4
Partition/aggregate application structure. (From Kurose, J.F. and Ross, K.W., *Computer Networking: A Top Down Approach*, 6th edn., Addison-Wesley, 2012.)

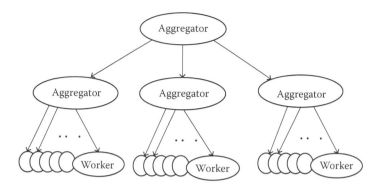

FIGURE 10.5
TCP incast.

mice collide [5]. Moreover, no worker node can transmit the next data block until all the worker nodes finish transmitting the current data block. Such a transmission is termed as barrier synchronized transmission [9].

Under such constraints, as the number of concurrent worker nodes increases, the perceived application level throughput at the aggregator decreases due to a large number of packet losses. The lost packets are retransmitted only after the retransmit timeout (RTO), which is generally in the order of a few *milliseconds*. As mentioned earlier, mice traffic requires short response time and is highly delay sensitive. Frequent timeouts resulting from incast significantly degrade the performance of mice traffic as the lost packets are retransmitted after a few *milliseconds*.

It must be noted that a *fast retransmit* mechanism may not be applicable to mice traffic applications since the data transmission volume of such traffic is quite less, and hence, there are very few packets in the entire flow. As a result, the sender (or worker node) may not get sufficient duplicate acknowledgements (*dupacks*) to trigger a fast retransmit.

Mitigating TCP incast: A lot of solutions, ranging from application layer solutions to transport layer solutions and link layer solutions, have been proposed recently to overcome the TCP incast problem. A few solutions suggest *revision of TCP*, others recommend to *replace TCP*, while some seek solutions from layers other than the transport layer to solve this problem [1]. Ren et al. [11] provides a detailed analysis and summary of all such solutions.

10.4.1.2 TCP Outcast

When a large set of flows and a small set of flows arrive at two different input ports of a switch and compete for the same bottleneck output port, the small set of flows lose out on their throughput share significantly. This phenomenon has been termed as TCP outcast [6] and mainly occurs in data center switches that employ drop-tail queues. Drop-tail queues lead to consecutive packet drops from one port and, hence, cause frequent TCP timeouts. This property

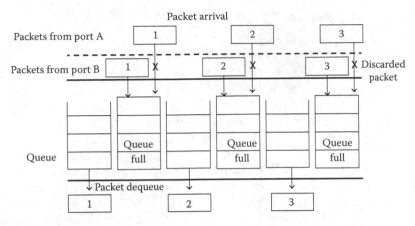

FIGURE 10.6
Example scenario of port blackout. (From Prakash, P. et al., The TCP outcast problem: Exposing unfairness in data center networks, in *Proceedings of the Ninth USENIX Conference on Networked Systems Design and Implementation*, ser. NSDI'12, USENIX Association, Berkeley, CA, 2012, pp. 30–30, available [Online]: http://dl.acm.org/citation.cfm?id=2228298.2228339.)

of drop-tail queues is termed as *port blackout* [6], and it significantly affects the performance of small flows because frequent timeouts lead to high latencies and, thus, poor-quality response times. Figure 10.6 shows an example scenario of a port blackout, where A and B are input ports and C is the common output port. The figure shows that packets arriving at port B are successfully queued whereas those arriving at port A are dropped consecutively.

It is well known that the throughput of a TCP flow is inversely proportional to the RTT of that flow. This behavior of TCP leads to RTT bias, that is, flows with low RTT achieve larger share of bandwidth than the flows with high RTT. However, it has been observed that due to TCP outcast problem in DCNs, TCP exhibits *inverse RTT bias* [6], that is, flows with low RTT are outcasted by flows with high RTT.

The two main factors that cause TCP outcast are (1) the usage of drop-tail queues in switches and (2) many-to-one communication pattern, which leads to a large set of flows and a small set of flows arriving at two different input ports and competing for the same bottleneck output port. Both these factors are quite common in DCNs since majority of the switches employ drop-tail queues and many-to-one communication pattern is the foundation of many cloud applications.

Mitigating TCP outcast: A straightforward approach to mitigate TCP outcast is to use queuing mechanisms other than drop tail, for example, random early detection (RED) [12] and stochastic fair queue (SFQ) [6]. Another possible approach is to minimize the buffer occupancy at the switches by designing efficient TCP congestion control laws at the end hosts.

10.4.1.3 Queue Buildup

Due to the diverse nature of cloud applications, mice traffic, cat traffic, and elephant traffic coexist in a DCN. The long-lasting and greedy nature of elephant traffic drives the network to the point of extreme congestion and overflows the bottleneck buffers. Thus, when both mice traffic and elephant traffic traverse through the same route, the performance of mice traffic is significantly affected due to the presence of the elephant traffic [5].

Following are two ways in which the performance of mice traffic is degraded due to the presence of elephant traffic [5]: (1) Since most of the buffer is occupied by elephant traffic, there is a high probability that the packets of mice traffic get dropped. The impact of this situation is similar to that of TCP incast because the performance of mice traffic is largely affected by frequent packet losses and, hence, the timeouts. (2) The packets of mice traffic, even when none are lost, suffer from increased queuing delay as they are in queue behind the packets of elephant traffic. This problem is termed as queue buildup.

Mitigating queue buildup: Queue buildup problem can be solved only by minimizing the queue occupancy in the DCN switches. Most of the existing TCP variants employ reactive approach toward congestion control, that is, they do not reduce the sending rate unless a packet loss is encountered and, hence, fail to minimize the queue occupancy. A proactive approach instead is desired to minimize the queue occupancy and overcome the problem of queue buildup.

10.4.1.4 Buffer Pressure

Buffer pressure is yet another impairment caused by the long-lasting and greedy nature of elephant traffic. When both mice traffic and elephant traffic coexist on the same route, most of the buffer space is occupied by packets from the elephant traffic. This leaves a very little room to accommodate the burst of mice traffic packets arising out of many-to-one communication pattern. The result is that a large number of packets from mice traffic are lost, leading to poor performance. Moreover, majority of the traffic in DCNs is bursty [5], and hence, packets of mice traffic get dropped frequently because the elephant traffic lasts for a longer time and keeps most of the buffer space occupied.

Mitigating buffer pressure: Like queue buildup, buffer pressure problem too can be solved by minimizing the buffer occupancy in the switches.

10.4.1.5 Pseudocongestion Effect

Virtualization is one of the key technologies driving the success of cloud computing applications. Modern data centers adopt virtual machines (VMs) to offer on-demand cloud services and remote access. These data centers are known as *virtualized data centers* [1,7]. Though there are several advantages of virtualization like efficient server utilization, service isolation, and low system maintenance cost [1], it significantly affects the environment where

TABLE 10.4

TCP Impairments in DCNs and Their Causes

TCP Impairment	Causes
TCP incast	Shallow buffers in switches and bursts of mice traffic resulting from many-to-one communication pattern
TCP outcast	Usage of tail-drop mechanism in switches
Queue buildup	Persistently full queues in switches due to elephant traffic
Buffer pressure	Persistently full queues in switches due to elephant traffic and bursty nature of mice traffic
Pseudocongestion effect	Hypervisor scheduling latency

our traditional protocols (e.g., TCP and UDP [user datagram protocol]) work. The recent study of Amazon EC2 data center reveals that virtualization dramatically deteriorates the performance of TCP and UDP in terms of both throughput and end-to-end delay [1]. Throughput becomes unstable, and the end-to-end delay becomes quite large even if the network load is less [1].

When more number of VMs are running on the same physical machine, the hypervisor scheduling latency increases the waiting time for each VM to obtain an access to the processor. Hypervisor scheduling latency varies from microseconds to several hundred milliseconds [7], leading to unpredictable network delays (i.e., RTT) and, thus, affecting the throughput stability and largely increasing the end-to-end delay. Moreover, hypervisor scheduling latency can be so high that it may lead to RTO at the VM sender. Once RTO occurs, VM sender assumes that the network is heavily congested and significantly brings down the sending rate. We term this effect as *pseudocongestion effect* because the congestion sensed by the VM sender is actually *pseudocongestion* [7].

Mitigating pseudocongestion effect: There are generally two possible approaches to address the aforementioned problem. One is to design efficient schedulers for hypervisor so that the scheduling latency can be minimized. Another approach is to modify TCP such that it can intelligently detect the *pseudocongestion* and react accordingly.

10.4.2 Summary: TCP Impairments and Causes

We briefly summarize the TCP impairments discussed in the previous subsections and mention the causes for the same in Table 10.4.

10.5 TCP Enhancements for DCNs

Recently, a few TCP variants, specifically data transport, have been proposed in DCNs. The major goal of these TCP variants is to overcome the aforementioned impairments and improve the performance of TCP in DCNs.

This chapter presents the background and the causes of each of the afore-mentioned impairments, followed by a comparative study of TCP variants that aim to overcome these impairments. Although a few other transport protocols have also been proposed for DCNs, we restrict the scope of this chapter to TCP variants because TCP is the most widely deployed transport protocol in modern operating systems.

10.5.1 TCP with Fine-Grained RTO (FG-RTO) [4]

The default value of minimum RTO in TCP is generally in the order of mil-liseconds (around 200 ms). This value of RTO is suitable for Internet-like sce-narios where the average RTT is in the order of hundreds of milliseconds. However, it is significantly larger than the average RTT in data centers, which is in the order of a few microseconds. A large number of packet losses due to TCP incast, TCP outcast, queue buildup, buffer pressure, and pseudoconges-tion effect result in frequent timeouts and, in turn, lead to missed deadlines and significant degradation in the performance of TCP. Vasudevan et al. [4] show that reducing the minimum RTO from 200 ms to 200 μs significantly alleviates the problems of TCP in DCNs and improves the overall through-put by several orders of magnitude.

Advantages: The major advantage of this approach is that it requires minimum modification to the traditional working of TCP and thus can be easily deployed.

Shortcomings: The real-time deployment of fine-grained timers is a challeng-ing issue because the present operating systems lack the high-resolution timers required for such low RTO values. Moreover, FG-RTOs may not be suitable for servers that communicate to clients through the Internet. Apart from the implementation issues of fine-grained timers, it must be noted that this approach of eliminating drawbacks of TCP in DCNs is a *reactive* approach. It tries to reduce the impact of a packet loss rather than *avoiding* the packet loss in the first place. Thus, although this approach significantly improves the network performance by reducing post–packet loss delay, it does not alleviate the TCP incast problem for loss-sensitive applications.

10.5.2 TCP with FG-RTO + Delayed ACKs Disabled [4]

Delayed ACKs are mainly used for reducing the overhead of ACKs on the reverse path. When delayed ACKs are enabled, the receiver sends only one ACK for every two data packets received. If only one packet is received, the receiver waits for delayed ACK timeout period before sending an ACK. This timeout period is usually 40 ms. This scenario may lead to spurious retransmissions if FG-RTO timers (as explained in the previous section) are deployed. The reason is that receiver waits for 40 ms before sending an ACK for the received packet and by that time, FG-RTO, which is in the order of a few microseconds, expires and forces the sender to retransmit the packet.

Thus, the delayed ACK timeout period either must be reduced to a few microseconds or must be completely disabled while using FG-RTOs to avoid such spurious retransmissions. This approach further enhances the TCP throughput in DCNs.

Advantages: It has been shown in [4] that reducing the delayed ACK timeout period to 200 μs while using FG-RTO achieves far better throughput than the throughput obtained when delayed ACKs are enabled. Moreover, completely disabling the delayed ACKs while using FG-RTO further improves the overall TCP throughput.

Shortcomings: The shortcomings of this approach are exactly similar to that of TCP with FG-RTO because this approach is an undesired side effect of the previous approach.

10.5.3 DCTCP [5]

Additive increase/multiplicative decrease (AIMD) is the cornerstone of TCP congestion control algorithms. When an acknowledgement (ACK) is received in AIMD phase, the congestion window (*cwnd*) is increased as shown in the following equation. This is known as additive increase phase of the AIMD algorithm:

$$cwnd = cwnd + \frac{1}{cwnd} \tag{10.1}$$

When congestion is detected either through *dupacks* or selective acknowledgement (SACK), *cwnd* is updated as shown in the following equation. This is known as multiplicative decrease phase of the AIMD algorithm:

$$cwnd = \frac{cwnd}{2} \tag{10.2}$$

Data center TCP (DCTCP) employs an efficient multiplicative decrease mechanism that reduces the *cwnd* based on the *amount of congestion* in the network rather than reducing it by half. DCTCP leverages explicit congestion notification (ECN) mechanism [13] to extract multibit feedback on the *amount of congestion* in the network from the single-bit stream of ECN marks. The next subsection describes the working of ECN in brief:

10.5.3.1 ECN

ECN [13] is one of the most popular congestion signaling mechanisms in communication networks. It is widely deployed in a large variety of operating systems at end hosts, modern Internet routers, and used by a variety of transport protocols. Moreover, it has been noticed that the use of ECN in the Internet has increased by threefolds in the last few years.

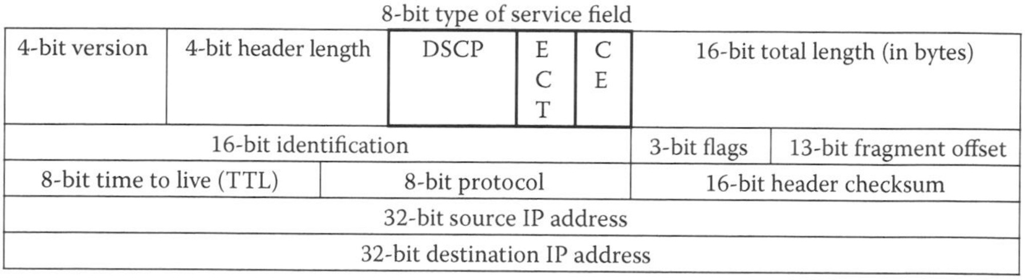

4-bit version	4-bit header length	8-bit type of service field				
		DSCP	E C T	C E	16-bit total length (in bytes)	
16-bit identification				3-bit flags	13-bit fragment offset	
8-bit time to live (TTL)		8-bit protocol		16-bit header checksum		
32-bit source IP address						
32-bit destination IP address						

FIGURE 10.7
ECN bits in IP header.

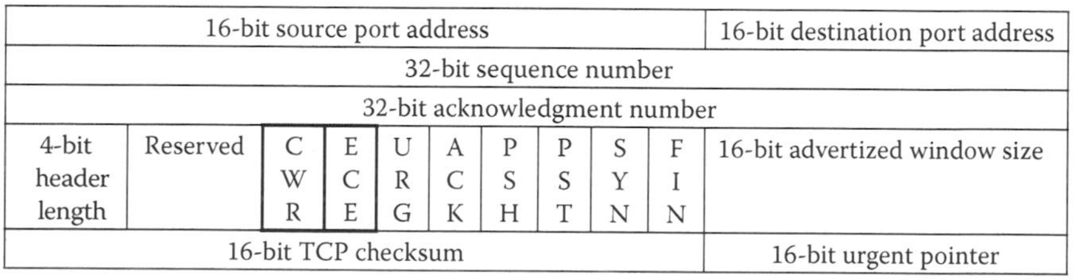

16-bit source port address									16-bit destination port address
32-bit sequence number									
32-bit acknowledgment number									
4-bit header length	Reserved	C W R	E C E	U R G	A C K	P S H	P S T	S Y N	F I N	16-bit advertized window size
16-bit TCP checksum									16-bit urgent pointer

FIGURE 10.8
ECN bits in TCP header.

As shown in Figures 10.7 and 10.8, ECN uses two bits in the IP header, namely ECN-capable transport (ECT) and congestion experienced (CE), and two bits in the TCP header, namely congestion window reduced (CWR) and ECN echo (ECE), for signaling congestion to the end hosts. ECN is an industry standard, and its detailed mechanism is described in RFC 3168. Tables 10.5 and 10.6 show the ECN codepoints in the TCP header and the IP header, respectively, and Figure 10.9 shows in brief the steps involved in the working of ECN mechanism.

As described in RFC 3168, the sender and the receiver must negotiate the use of ECN during the three-way handshake (see Figure 10.10). If both are ECN capable, the sender marks every outgoing data packet with either ECT(1) codepoint or ECT(0) codepoint. This serves as an indication to the router that both sender and receiver are ECN capable. Whenever congestion builds up, the router marks the data packet by replacing ECT(1) or ECT(0) codepoint

TABLE 10.5

ECN Codepoints in the TCP Header

Codepoint	CWR Bit Value	ECE Bit Value
Non–ECN setup	0	0
ECN echo	0	1
CWR	1	0
ECN setup	1	1

TABLE 10.6

ECN Codepoints in the IP Header

Codepoint	ECT Bit Value	CE Bit Value
Non-ECT	0	0
ECT(1)	0	1
ECT(0)	1	0
CE	1	1

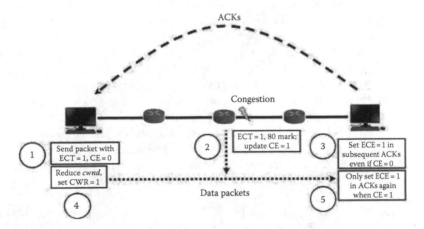

FIGURE 10.9
ECN.

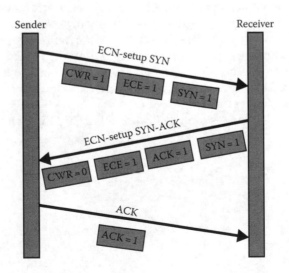

FIGURE 10.10
ECN negotiation.

by the CE codepoint. When the receiver receives a marked packet with CE codepoint, it infers congestion and hence marks *a series of* outgoing acknowledgments (ACKs) with ECE codepoint until the sender acknowledges with CWR codepoint (see Figure 10.9).

The major observation here is that, even if the router marks just one data packet, the receiver continues to mark ACKs with ECE until it receives confirmation from the sender (see Step 3 of Figure 10.9). This is to ensure the reliability of congestion notification, because even if the first marked ACK is lost, other marked ACKs would notify the sender about congestion. Note that this basic working of ECN aims to only notify the sender about congestion. It is not designed to provide the additional information about the *amount of congestion* to the sender.

At the receiver, counting the number of packets marked by the router provides fairly accurate information about the *amount of congestion* in the network. However, conveying this information to the sender by using ECN is a complex task. One of the possible ways is to enable the sender to count the number of marked ACKs it receives from the receiver. The limitation, however, is that even if router marks just one data packet, the receiver sends *a series of* marked ACKs. Hence, the number of marked ACKs counted by the sender would be much higher than the number of packets actually marked by the router. This would lead to incorrect estimation of the *amount of congestion* in the network.

To overcome this limitation, DCTCP modifies the basic mechanism of ECN. Unlike TCP receiver, which sends *a series of* marked ACKs, DCTCP receiver sends a marked ACK *only when* it receives a marked packet from the router, that is, it sets ECE codepoint in the outgoing ACK *only when* it receives a packet with CE codepoint. Thus, the DCTCP sender obtains an accurate number of packets marked by the router by simply counting the number of marked ACKs it receives. Note that this modification to the original ECN mechanism reduces the reliability because if a marked ACK is lost, sender remains unaware of the congestion and does not reduce the sending rate. However, since DCNs are privately controlled networks, the possibility that an ACK gets lost is negligible.

On receiving the congestion notification via ECN, the *cwnd* in DCTCP is reduced as shown in the following:

$$cwnd = cwnd \times \left(1 - \frac{\alpha}{2}\right) \tag{10.3}$$

where $\alpha (0 < \alpha \leq 1)$ is an estimate of the fraction of packets that are marked and is calculated as shown in (10.4). F in (10.4) is the fraction of packets that are marked in the previous *cwnd* and g ($0 < g < 1$) is the exponential weighted moving average constant. Thus, when congestion is low (α *is near* 0), *cwnd* is reduced slightly and when congestion is high (α *is near* 1), *cwnd* is reduced by half, just like the traditional TCP:

$$\alpha = (1-g) \times \alpha + g \times F \qquad (10.4)$$

The major goal of DCTCP algorithm is to achieve low latency (desirable for mice traffic), high throughput (desirable for elephant traffic), and high burst tolerance (to avoid packet losses due to incast). DCTCP achieves these goals by reacting to the *amount of congestion* rather than halving the *cwnd*. DCTCP uses a marking scheme at switches that sets the CE codepoint [13] of packets as soon as the buffer occupancy exceeds a fixed predetermined threshold, K (17% as mentioned in [14]). The DCTCP source reacts by reducing the window by a factor that depends on the fraction of marked packets: the larger the fraction, the bigger the decrease factor.

Advantages: DCTCP is a novel TCP variant that alleviates TCP incast, queue buildup, and buffer pressure problems in DCNs. It requires minor modifications to the original design of TCP and ECN to achieve these performance benefits. DCTCP employs a *proactive* behavior, that is, it tries to avoid packet loss. It has been shown in [5] that when DCTCP uses FG-RTO, it further reduces the impact of TCP incast and also improves the scalability of DCTCP. The stability, convergence, and fairness properties of DCTCP [14] make it a suitable solution for implementation in DCNs. Moreover, DCTCP is already implemented in the latest versions of Microsoft Windows server operating system.

Shortcomings: The performance of DCTCP falls back to that of TCP when the degree of incast increases beyond 35, that is, if there are more than 35 worker nodes sending data to the same aggregator, DCTCP fails to avoid incast and its performance falls back to that of the traditional TCP. However, authors show that dynamic buffer allocation at the switch and usage of FG-RTO can scale DCTCP's performance to handle up to 40 worker nodes in parallel.

Although DCTCP uses a simple queue management mechanism at the switch, it is ambiguous whether DCTCP can alleviate the problem of TCP outcast. Similarly, DCTCP does not address the problem of pseudocongestion effect in virtualized data centers. DCTCP utilizes minimum buffer space in the switches, which, in fact, is a desirable property to avoid TCP outcast. However, experimental studies are required to conclude whether DCTCP can mitigate the problems of TCP outcast and pseudocongestion effect.

10.5.4 ICTCP [9]

Like DCTCP, the main idea of incast congestion control for TCP (ICTCP) is to *avoid* packet losses due to congestion rather than to avoid from the packet losses. It is well known that a TCP sender can send a minimum of advertised window (*rwnd*) and congestion window (*cwnd*) (i.e., min (*rwnd*, *cwnd*)). ICTCP leverages this property and efficiently varies the *rwnd* to avoid TCP incast. The major contributions of ICTCP are the following: (1) The available bandwidth is used as a quota to coordinate the *rwnd* increase of all connections, (2) per-flow congestion control is performed independently, and (3) *rwnd* is adjusted based on the ratio of difference between expected throughput and measured throughput over expected throughput. Moreover, live RTT is used for the throughput estimation.

Advantages: Unlike DCTCP, ICTCP does not require any modifications at the sender side (i.e., worker nodes) or network elements such as routers and switches. Instead, ICTCP requires modification only at the receiver side (i.e., an aggregator). This approach is adopted to retain the backward compatibility and make the algorithm general enough to handle the incast congestion in future high-bandwidth, low-latency networks.

Shortcomings: Although it has been shown in [9] that ICTCP achieves almost zero timeouts and high throughput, the scalability of ICTCP is a major concern, that is, the capability to handle incast congestion when there are extremely large number of worker nodes since ICTCP employs per-flow congestion control. Another limitation of ICTCP is that it assumes that both the sender and the receiver are under the same switch, which might not be always the case. Moreover, it is not known how much buffer space is utilized by ICTCP. Thus, it is difficult to conclude whether ICTCP can alleviate queue buildup, buffer pressure, and TCP outcast problems. Like DCTCP, ICTCP too does not address the problem of pseudocongestion effect in virtualized data centers.

10.5.5 IA-TCP [15]

Unlike DCTCP and ICTCP that use window-based congestion control, incast avoidance algorithm for TCP (IA-TCP) uses a rate-based congestion control algorithm to control the total number of packets injected in the network. The motivation behind selecting rate-based congestion control mechanism is that window-based congestion control mechanisms in DCNs have limitations in terms of scalability, that is, number of senders in parallel.

The main idea of IA-TCP is to limit the total number of outstanding data packets in the network so that it does not exceed the bandwidth-delay product (BDP). IA-TCP employs ACK regulation at the receiver and, like ICTCP, leverages the advertised window (*rwnd*) field of the TCP header to regulate the *cwnd* of every worker node. The minimum *rwnd* is set to 1 packet. However, if a large number of worker nodes send packets with respect to a minimum *rwnd* of 1 packet, the total number of outstanding packets in the network may exceed the link capacity. In such scenarios, IA-TCP adds delay, Δ, to the ACK packet to ensure that the aggregate data rate does not exceed the link capacity. Moreover, IA-TCP also uses delay, Δ, to avoid the synchronization among the worker nodes while sending the data.

Advantages: Like ICTCP, IA-TCP also requires modification only at the receiver side (i.e., an aggregator) and does not require any modifications at the sender or network elements. IA-TCP achieves high throughput and significantly improves the query completion time. Moreover, the scalability of IA-TCP is clearly demonstrated by configuring up to 96 worker nodes sending data in parallel.

Shortcomings: Similar to the problem of ICTCP, it is not clear how much buffer space is utilized by IA-TCP. Thus, experimental studies are required to conclude whether IA-TCP can mitigate queue buildup, buffer pressure, and TCP outcast problems. Like DCTCP and ICTCP, studies are required in virtualized data center environments to analyze the performance of IA-TCP with respect to the problem of pseudocongestion effect.

10.5.6 D²TCP [16]

Deadline-aware data center TCP (D²TCP) is a novel TCP-based transport protocol that is specifically designed to handle high burst situations. Unlike other TCP variants that are deadline agnostic, D²TCP is deadline aware. D²TCP uses a *distributed* and *reactive* approach for bandwidth allocation and employs a novel deadline-aware *congestion avoidance* algorithm that uses ECN feedback and deadlines to vary the sender's *cwnd* via a gamma-correction function [16].

D²TCP does not maintain per-flow information and, instead, inherits the distributed and reactive nature of TCP while adding deadline awareness to it. Similarly, D²TCP employs its congestion avoidance algorithm by adding deadline awareness to DCTCP. The main idea, thus, is that far-deadline flows back off aggressively and the near-deadline flows back off only a little or not at all.

Advantages: The novelty of D²TCP lies in the fact that it is deadline aware and reduces the fraction of missed deadlines up to 75% as compared to DCTCP. In addition, since it is designed upon DCTCP, it avoids TCP incast and queue buildup and has high burst tolerance.

Shortcomings: The shortcomings of D²TCP are exactly similar to those of DCTCP: scalability and whether it is robust against TCP outcast as well as pseudocongestion effect. However, since it is deadline aware, it would be interesting to analyze the robustness of D²TCP against the pseudocongestion effect in virtualized data centers.

10.5.7 TCP-FITDC [17]

TCP-FITDC is an adaptive delay-based mechanism to *prevent* the problem of TCP incast. Like D²TCP, TCP-FITDC is also a DCTCP-based TCP variant that benefits from the novel ideas of DCTCP. Apart from utilizing ECN as an indicator of network buffer occupancy and buffer overflow, TCP-FITDC also monitors changes in the queueing delay to estimate variations in the available bandwidth.

If there is no marked ACK received during the RTT, TCP-FITDC implies the queue length in the switch is below the marking threshold, and hence, TCP-FITDC increases the *cwnd* to improve the throughput. If marked ACKs are received during the RTT, *cwnd* is decreased to control the queue length. TCP-FITDC maintains two separate variables called rtt_1 and rtt_2 for unmarked ACKs and marked ACKs, respectively. By analyzing the difference between these two types of ACKs, TCP-FITDC gets more accurate estimate of the network conditions. The *cwnd* is then reasonably decreased to maintain low queue length.

Advantages: TCP-FITDC gets a better estimate of the network conditions by coupling the information received via ECN and the information obtained by monitoring the RTT. Thus, it has better scalability than DCTCP and scales up to 45 worker nodes in parallel. It avoids TCP incast and queue buildup and has a high burst tolerance because it is built upon DCTCP.

Shortcomings: The shortcomings of TCP-FITDC are similar to those of DCTCP, except that it improves the scalability of DCTCP. Unlike D^2TCP, TCP-FITDC is deadline agnostic, and like all the aforementioned TCP variants, it does not address TCP outcast and pseudocongestion effect problems.

10.5.8 TDCTCP [18]

TDCTCP attempts to improvise the working of DCTCP (thus, it is DCTCP based) by making three modifications. First, unlike DCTCP, TDCTCP not only decreases but also increases the *cwnd* based on the *amount of congestion* in the network, that is, instead of increasing the *cwnd* as shown in (10.1), TDCTCP increases the *cwnd* as shown in (10.5). Thus, when the network is lightly loaded, the increment in *cwnd* is high, and vice versa:

$$cwnd = cwnd \times \left(1 + \frac{1}{1 + (\alpha/2)} \right) \tag{10.5}$$

Second, TDCTCP resets the value of α after every delayed ACK timeout. This is done to ensure that α does not carry the stale information about the network conditions, because if the stale value of α is high, it restricts the *cwnd* increment and thereby degrades the overall throughput. Lastly, TDCTCP employs an efficient approach to dynamically calculate the delayed ACK timeout with a goal to achieve better fairness.

Advantages: TDCTCP achieves 26%–37% better throughput and 15%–20% better fairness than DCTCP in a wide variety of scenarios ranging from single bottleneck topologies to multibottleneck topologies and varying buffer sizes. Moreover, it achieves better throughput and fairness even at very low values of *K*, that is, the ECN marking threshold at the switch. However, there is a slight increase in the delay and queue length while using TDCTCP as compared to DCTCP.

Shortcomings: Although TDCTCP improves throughput and fairness, it does not address the scalability challenges faced by DCTCP. Like most of the other TCP variants discussed, TDCTCP too is deadline agnostic and does not alleviate the problems of TCP outcast and pseudocongestion effect.

10.5.9 TCP with Guarantee Important Packets (GIP) [19]

TCP with GIP mainly aims to improve the network performance in terms of goodput by minimizing the total number of timeouts. Timeouts lead to a dramatic degradation in the network performance and affect the user perceived delay. The authors of TCP with GIP focus on avoiding mainly two types of timeouts in the network: (1) the timeouts due to the loss of full window of packets, full window loss timeouts (FLoss-TOs), and (2) the timeouts due to the lack of ACKs, lack of ACKs timeouts (LAck-TOs).

FLoss-TOs generally occur when the total data sent by all the worker nodes exceed the available bandwidth in the network, and thus, a few unlucky flows end up losing all the packets of the window. On the other hand, LAck-TOs mainly occur when the transmission is *barrier-synchronized transmission*. In such transmissions, the aggregator will not request the worker nodes to transmit the next stripe units until all the worker nodes finish sending their current ones. If a few packets get dropped at the end of the stripe unit, they cannot be recovered until the RTO fires because there may not be sufficient *dupacks* to trigger fast retransmit.

TCP with GIP introduces *flags* in the interface between the application layer and the transport layer. These *flags* indicate whether the running application follows many-to-one communication pattern or not. If the running application follows such a communication pattern, TCP with GIP redundantly transmits the last packet of the stripe unit at most three times and each worker node decreases its initial *cwnd* at the head of the stripe unit. On the other hand, if the running application does not follow many-to-one communication pattern, TCP with GIP works like a standard TCP.

Advantages: TCP with GIP achieves almost zero timeouts and higher goodput in a wide variety of scenarios including with and without the background UDP traffic. Moreover, the scalability of TCP with GIP is much more than any other TCP variant discussed earlier, that is, it scales well up to 150 worker nodes in parallel.

Shortcomings: TCP with GIP does not address the queue occupancy problem resulting from the presence of elephant traffic. As a result, the queue buildup, buffer pressure, and TCP outcast problems remain unsolved because all these problems arise due to the lack of the buffer space in the switches. Although timeouts are eliminated by TCP with GIP, flows may miss the specified deadlines because of queueing delay. Moreover, the hypervisor scheduling latency is not taken into consideration, and thus, the problem of pseudocongestion effect also remains open. Note that high latencies introduced by hypervisor scheduling algorithm may also prevent flows from meeting the specified deadlines.

10.5.10 PVTCP [7]

Paravirtualized TCP (PVTCP) proposes an efficient solution to the problem of pseudocongestion effect. This approach does not require any changes to be done in the hypervisor. Instead, the basic working of TCP is modified to accept the latencies introduced by the hypervisor scheduler. An efficient approach is suggested to capture the *actual* picture of every packet transmission involving the hypervisor-introduced latencies and then determine RTO more accurately to filter out pseudocongestion effect.

Whenever the hypervisor introduces scheduling latency, sudden spikes can be observed during the regular measurements of RTT. PVTCP detects these sudden spikes and filters out the negative impact of these spikes by proper RTT measurement and RTO management. While calculating average RTT, PVTCP ignores the measurement of a particular RTT if a spike is observed in that RTT.

Advantages: PVTCP solves the problem of pseudocongestion effect without requiring any changes in the hypervisor. By detecting the unusual spikes, accurately measuring RTT, and properly managing RTO, PVTCP enhances the performance of virtualized data centers.

Shortcomings: The scalability of PVTCP is ambiguous, and thus, whether it can solve TCP incast effectively or not is unclear. The queue occupancy while using PVTCP is not taken into consideration, which may further lead to problems such as queue buildup, buffer pressure, TCP outcast, and missed deadlines.

10.5.11 Summary: TCP Enhancements for DCNs

Table 10.7 summarizes the comparative study of TCP variants proposed for DCNs. Apart from the novelty of the proposed TCP variant, the table also highlights the deployment complexity of each protocol. The protocols that require modifications in the sender, receiver, and switch are considered as hard to deploy. The ones that require modification only at the sender or receiver are considered as easy to deploy. DCNs, however, are privately controlled and managed networks, and thus, the former ones may also be treated as easy to deploy.

Apart from the aforementioned parameters, the summary also includes which problems among TCP incast, TCP outcast, queue buildup, buffer pressure, and pseudocongestion effect are alleviated by each TCP variant. The details regarding the tools used/approach of implementation adopted by the authors are also listed.

TABLE 10.7

Summary of TCP Variants Proposed for DCNs

TCP Variants Proposed for DCNs	Modifies Sender	Modifies Receiver	Modifies Switch	Solves TCP Incast	Solves TCP Outcast	Solves Queue Buildup	Solves Buffer Pressure	Is Deadline Aware?	Detects Pseudocongestion	Implementation
TCP with FG-RTO	✓	×	×	✓	×	✗	×	×	×	Testbed and ns-2
TCP with FG-RTO + delayed ACKs disabled	✓	×	×	✓	×	✗	×	×	×	Testbed and ns-2
DCTCP	✓	✓	✓	✓	×	✓	✓	×	×	Testbed and ns-2
ICTCP	×	✓	×	✓	×	×	✗	×	×	Testbed
IA-TCP	×	✓	×	✓	×	×	×	×	×	ns-2
D²TCP	✓	✓	✓	✓	×	✓	✓	✓	×	Testbed and ns-3
TCP-FITDC	✓	✓	✓	✓	×	✓	✓	×	×	Modeling and ns-2
TDCTCP	✓	✓	✓	✓	×	✓	✓	×	×	OMNeT++
TCP with GIP	×	✓	×	✓	×	×	×	×	×	Testbed and ns-2
PVTCP	✓	✓	×	✓	×	×	×	×	✓	Testbed

Although several modifications have been proposed to the original design of TCP, there is an acute need to further optimize the performance of TCP in DCNs. A few open issues are listed in the following:

1. Except D²TCP, all other TCP variants are deadline agnostic. Meeting deadlines is the most important requirement in DCNs. Missed deadlines may lead to violations of SLAs and, thus, incur high cost to the organization.

2. Most of the data centers today employ virtualization for efficient resource utilization. Hypervisor scheduling latency ranges from microseconds to hundreds of milliseconds and, hence, may hinder in successful completion of flows within the specified deadline. While making modifications to hypervisors is one viable solution, designing an efficient TCP that is deadline aware and automatically tolerates hypervisor scheduling latency is a preferred solution.

3. A convincing solution to TCP outcast problem is unavailable. An optimal solution to overcome TCP outcast must ensure minimal buffer occupancy at the switch. Since RED is implemented in most of the modern switches, it can be used to control the buffer occupancy. The parameter sensitivity of RED, however, poses further challenges and complicates the problem.

10.6 Summary

Data centers in the present scenario house a plethora of Internet applications. These applications are diverse in nature and have various performance requirements. Understanding the complete architecture of a data center from the point of its physical as well as networking infrastructure is crucial to the success of cloud computing.

Although several issues related to networking still exist in data centers, one of the most crucial ones is to meet the diverse requirements of various traffic types that coexist in a data center environment. Majority of the traffic uses many-to-one communication pattern to gain performance efficiency. TCP, which has been a mature transport protocol of Internet since the past several decades, suffers from performance impairments such as TCP incast, TCP outcast, queue buildup, buffer pressure, and pseudocongestion effect in DCNs. We described each of the aforementioned impairment in brief along with the causes and possible approaches to mitigate them. A comparative study of TCP variants that have been specifically designed for DCNs is presented, while describing the advantages and shortcomings of each.

Review Points

- Data centers are mainly classified based on the uptime they provide to the cloud user (see Section 10.1).
- Around 15% of the cost in data centers goes in networking (see Section 10.3).
- Different traffic types with diverse performance requirement coexist in DCNs (see Section 10.2.6).
- Traditional TCP/IP does not provide optimal network performance in DCNs (see Section 10.4).
- Although a few new TCP algorithms have been designed to enhance the network performance in data centers, a lot remains to be done (see Section 10.5).

Review Questions

1. What are the different ways to classify data centers?
2. Explain the different types of traffic in DCNs. Mention their performance requirements and provide an example for each type of traffic.
3. How is queue buildup different from buffer pressure?
4. Describe at least two approaches to solve the problem of pseudocongestion in virtualized data centers.
5. Why traditional ECN mechanism cannot be used in DCTCP?

References

1. Zhang, J., F. Ren, and C. Lin. Survey on transport control in data center networks. *IEEE Network* 27(4): 22–26, 2013.
2. Greenberg, A., J. Hamilton, D. A. Maltz, and P. Patel. The cost of a cloud: Research problems in data center networks. *SIGCOMM Computer Communication Review* 39(1): 68–73, December 2008. Available [Online]: http://doi.acm.org/10.1145/1496091.1496103.
3. Kant, K. Data center evolution. *Computer Networks* 53(17): 2939–2965, December 2009. Available [Online]: http://dx.doi.org/10.1016/j.comnet.2009.10.004.

4. Vasudevan, V., A. Phanishayee, H. Shah, E. Krevat, D. G. Andersen, G. R. Ganger, G. A. Gibson, and B. Mueller. Safe and effective fine-grained TCP retransmissions for datacenter communication. *SIGCOMM Computer Communications Review* 39(4): 303–314, August 2009. Available [Online]: http://doi.acm.org/10.1145/1594977.1592604.

5. Alizadeh, M., A. Greenberg, D. A. Maltz, J. Padhye, P. Patel, B. Prabhakar, S. Sengupta, and M. Sridharan. Data center TCP (DCTCP). *SIGCOMM Computer Communications Review* 40(4): 63–74, August 2010. Available [Online]: http://doi.acm.org/10.1145/1851275.1851192.

6. Prakash, P., A. Dixit, Y. C. Hu, and R. Kompella. The TCP outcast problem: Exposing unfairness in data center networks. *Proceedings of the Ninth USENIX Conference on Networked Systems Design and Implementation*, ser. NSDI'12. USENIX Association, Berkeley, CA, 2012, pp. 30–30. Available [Online]: http://dl.acm.org/citation.cfm?id=2228298.2228339.

7. Cheng, L., C.-L. Wang, and F. C. M. Lau. PVTCP: Towards practical and effective congestion control in virtualized datacenters. *21st IEEE International Conference on Network Protocols*, ser. ICNP 2013. IEEE, 2013.

8. Chen, Y., R. Griffith, J. Liu, R. H. Katz, and A. D. Joseph. Understanding TCP incast throughput collapse in datacenter networks. *Proceedings of the First ACM Workshop on Research on Enterprise Networking*, ser. WREN'09. ACM, New York, 2009, pp. 73–82. Available [Online]: http://doi.acm.org/10.1145/1592681.1592693.

9. Wu, H., Z. Feng, C. Guo, and Y. Zhang. ICTCP: Incast congestion control for TCP in data center networks. *Proceedings of the Sixth International Conference*, ser. Co-NEXT'10. ACM, New York, 2010, pp. 13:1–13:12. Available [Online]: http://doi.acm.org/10.1145/1921168.1921186.

10. Kurose, J. F. and K. W. Ross. *Computer Networking: A Top Down Approach*, 6th edn. Addison-Wesley, 2012.

11. Ren, Y., Y. Zhao, P. Liu, K. Dou, and J. Li. A survey on TCP incast in data center networks. *International Journal of Communication Systems* pp. n/a–n/a, 2012. Available [Online]: http://dx.doi.org/10.1002/dac.2402.

12. Floyd, S. and V. Jacobson. Random early detection gateways for congestion avoidance. *IEEE/ACM Transactions on Networking* 1: 397–413, August 1993. Available [Online]: http://dx.doi.org/10.1109/90.251892.

13. Ramakrishnan, K. K. and S. Floyd. The addition of explicit congestion notification (ECN) to IP, 2001, rFC 3168.

14. Alizadeh, M., A. Javanmard, and B. Prabhakar. Analysis of DCTCP: Stability, convergence and fairness. *Proceedings of the ACM SIGMETRICS, Joint International Conference on Measurement and Modeling of Computer Systems*, ser. SIGMETRICS'11. ACM, New York, 2011, pp. 73–84. Available [Online]: http://doi.acm.org/10.1145/1993744.1993753.

15. Hwang, J., J. Yoo, and N. Choi. IA-TCP: A rate based incast-avoidance algorithm for TCP in data center networks. *IEEE ICC 2012*, Ottawa, Canada, 2012.

16. Vamanan, B., J. Hasan, and T. Vijaykumar. Deadline-aware datacenter TCP (D2TCP). *SIGCOMM Computer Communications Review* 42(4): 115–126, August 2012. Available [Online]: http://doi.acm.org/10.1145/2377677.2377709.

17. Wen, J., W. Zhao, J. Zhang, and J. Wang. TCP-FITDC: An adaptive approach to TCP incast avoidance for data center applications. *Proceedings of the 2013 International Conference on Computing, Networking and Communications (ICNC)*, ser. *ICNC'13*. IEEE Computer Society, Washington, DC, 2013, pp. 1048–1052. Available [Online]: http://dx.doi.org/10.1109/ICCNC.2013.6504236.

18. Das, T. and K. M. Sivalingam. TCP improvements for data center networks. *2013 Fifth International Conference on Communication Systems and Networks (COMSNETS)*. IEEE, 2013, pp. 1–10.

19. Zhang, J., F. Ren, L. Tang, and C. Lin. Taming TCP incast throughput collapse in data center networks. *21st IEEE International Conference on Network Protocols*, ser. *ICNP 2013*. IEEE, 2013.

11

Cloud Service Providers

Learning Objectives

The main objective of this chapter is to provide an overview of different cloud service providers. After reading this chapter, you will

- Know about different companies that support cloud computing
- Understand open source/proprietary tools offered by the companies
- Know cloud services offered by the companies
- Understand the features and available architecture of different tools

Preamble

This chapter provides an overview of cloud services offered by various companies. We begin with the introduction to cloud services. Subsequent sections talk about companies like Amazon, Microsoft, Google, EMC, Salesforce, and IBM that provide various tools and services in order to give cloud support. Each section briefly describes cloud features supported by these companies. It also gives an idea of tools and technologies adapted by companies in order to provide services to the users. In this chapter, we focus on giving readers brief information about various tools and technologies provided by different companies. After reading this chapter, the reader will be able to distinguish different services provided by various companies and make appropriate choices as per the requirement.

11.1 Introduction

Cloud computing is one of the most popular buzzwords used these days. It is the upcoming technology provisioning resources to the consumers in the form of different services like software, infrastructure, platform, and security. Services are made available to users on demand via the Internet from a cloud computing provider's servers as opposed to being provided from a company's own on-premise servers. Cloud services are designed to provide easy, scalable access to applications, resources, and services and are fully managed by a cloud service provider. A cloud service can dynamically scale to meet the needs of its users, and because the service provider supplies the hardware and software necessary for the service, there is no need for a company to provision or deploy its own resources or allocate information technology (IT) staff to manage the service. Examples of cloud services include online data storage and backup solutions, web-based e-mail services, hosted office suites and document collaboration services, database processing, and managed technical support services.

Cloud services can be broadly classified into three types: Software as a Service (SaaS), Platform as a Service (PaaS), and Infrastructure as a Service (IaaS). With growing technologies, many more services are emerging in this field, such as Security as a Service (SeaaS), Knowledge as a Service, and Data Analytics as a Service.

Many companies have come forward to adapt the cloud environment and ensure that the users as well as the companies benefit from this. Amazon, Microsoft, Google, Yahoo, EMC, Salesforce, Oracle, IBM, and many more companies provide various tools and services in order to give cloud support for their customers.

11.2 EMC

EMC is one of the leading global enterprises that require dynamic scalability and infrastructure agility to meet changing applications as well as business needs. EMC chose cloud computing as the ideal solution to reduce the complexity and optimize the infrastructure. Offering Information Technology as a Service (ITaaS) reduces the energy consumption through resource sharing.

11.2.1 EMC IT

Virtualization is the main concept behind the success of EMC IT. By virtualizing the infrastructure, allocation of the resources on demand is possible. This also helps to increase efficiency and resource utilization.

EMC IT provides its business process units with IaaS, PaaS, and SaaS. Figure 11.1 gives an overview of the services offered by EMC, which are explained in the following:

1. IaaS offers EMC business units the ability to provision infrastructure components such as network, storage, computing, and operating systems individually or as integrated services.

2. PaaS provides the secure application and information frameworks on top of application server, web server, database, unstructured content management, and security components as a service to business units from which to develop solutions. EMC IT offers database platforms (Oracle Database as a Service, SQL Server as a Service, Greenplum as a Service) and application platforms (application development, Enterprise Content Management as a Service, Information Cycle Management as a Service, Security PaaS, Integration as a Service) for the purpose of development.

3. SaaS provides applications and tools in a service model for business enablement. EMC IT brought together several existing business solutions under the unified architecture named as Business Intelligence as a Service. It also offers Enterprise Resource Planning (ERP) and Customer Relationship Management (CRM) as a Service.

4. User Interface as a Service (UIaaS) provisions user and interface experience, rather than provisioning the actual device used.

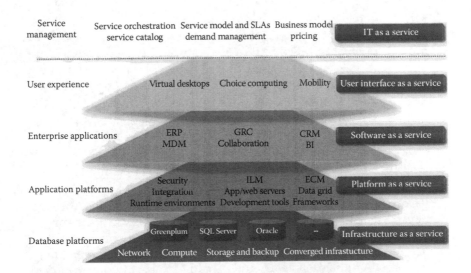

FIGURE 11.1
Cloud services by EMC. (Adapted from EMC IT's journey to the private cloud, applications and the cloud experience, White Paper-EMC.)

11.2.2 Captiva Cloud Toolkit

EMC offers a tool called *Captiva Cloud Toolkit* to help in the development of softwares. EMC Captiva Cloud Toolkit is a Software Development Kit (SDK) comprised of modules that help web application developers to quickly add scanning and imaging functionality directly to their web-based business applications. It is ideal for document capture vendors, commercial software developers, and enterprises that want to create custom web-based applications that are fully scan enabled, complimenting their business solution offerings.

Using Captiva Cloud Toolkit, developers can quickly create a working scan-enabled web-based business application in as early as 1 week. As a result, time to market is shortened and development, testing, and support costs are greatly reduced. Also, the enterprise's return on investment is quickly achieved, and its ability to compete in an increasingly competitive distributed document capture market is accelerated.

There are a few modules that are commonly used in most of the process development. These are basic modules that import images from various sources like fax, e-mail, or scanner or from any repository. A few of these modules are as follows:

1. *Scan*: Scanning is importing activity of documents into Captiva from a scanner. Basically, scanning happens at page level to bring images page by page into Captiva. Scanning is the entry point to Captiva where one can import any kind of document like pdf, tiff, and jpg.

2. *MDW*: Multi Directory Watch is another entry point to Captiva. MDW can be pointed to any folder/repository from where Captiva could import documents directly. MDW is very useful if business is getting documents in the form of a soft copy, for example, as an attached file in an e-mail. MDW also acts as a scan module except it does not interlock with the scanner.

3. *IE*: Image enhancement is a kind of filter or repairing tool for images that are not clear. It enhances the image quality, so it could be processed easily through Captiva. One can configure IE as per business requirement and images being received. The functionalities of IE are deskew, noise removal, etc.

4. *Index*: Indexing is a data capturing activity in Captiva through which one can capture key data from various fields. For example, if bank form is being processed, the A/C no. and sort code could be the indexing field. Indexing could be added as per requirement of business. A validation field could be added to avoid unwanted data entry while indexing any document.

5. *Export*: Export is the exit point of Captiva where images/data are sent to various repositories like file, net, document, or data. The exported data are used for business requirements of various business divisions. For example, if we are capturing the A/C no. and sort code for a bank application, this could be mapped to any department where it is needed.

6. *Multi*: Multi is the last process in Captiva to delete batches that have gone through all modules and exported value successfully. Multi could be configured as per need of business. In the case when it is required to take a backup of batches, this module could be avoided.

The previously mentioned modules are very basic modules of Captiva for indexing and exporting. But for more flexibility and automation, dispatcher is used, which is more accurate to capture data.

11.3 Google

Google is one among the leading cloud providers that offer secure storage of user's data. It provides cloud platform, app engine, cloud print, cloud connect, and many more features that are scalable, reliable, as well as secure. Google offers many of these services for free or at a minimum cost making it user friendly.

11.3.1 Cloud Platform

Google Cloud Platform enables developers to build, test, and deploy applications on Google's highly scalable and reliable infrastructure. Google has one of the largest and most advanced networks across the globe. Software infrastructures such as MapReduce, BigTable, and Dremel are the innovations for industrial development.

Google Cloud Platform includes virtual machines, block storage, NoSQL datastore, and big data analytics. It provides a range of storage services that allow easy maintenance and quick access of user's data. The cloud platform offers a fully managed platform as well as flexible virtual machines allowing the user to choose as per the requirements. Google also provides easy integration of user's application within the cloud platform.

Applications hosted on the cloud platform can automatically scale up to handle the most demanding workloads and scale down when traffic subsides. The cloud platform is designed to scale like Google's own products, even when there is a huge traffic spike. Managed services such as App Engine or Cloud Datastore provide autoscaling that enables application to grow with the users. The user has to pay only for what he or she uses.

11.3.2 Cloud Storage

Google Cloud Storage is a RESTful online file storage web service for storing and accessing one's data on Google's infrastructure. Representational state transfer (REST) is an architectural style consisting of a coordinated set of architectural constraints applied to components, connectors, and data elements within a distributed system. The service combines the performance and scalability of Google's cloud with advanced security and sharing capabilities. Google Cloud Storage is safe and secure. Data are protected through redundant storage at multiple physical locations.

The following are the few tools for Google Cloud Storage:

- *Google Developers Console* is a web application where one can perform simple storage management tasks on the Google Cloud Storage system.
- *gsutil* is a Python application that lets the user access Google Cloud Storage from the command line.

11.3.3 Google Cloud Connect

Google Cloud Connect is a feature provided by Google Cloud by integrating cloud and the application programming interface (API) for Microsoft Office. After installing a plug-in for the Microsoft Office suite of programs, one can save files to the cloud. The cloud copy of the file becomes the master document that everyone uses. Google Cloud Connect assigns each file a unique URL that can be shared to let others view the document.

If changes are made to the document, those changes will show up for everyone else viewing it. When multiple people make changes to the same section of a document, Cloud Connect gives chance to the user to choose which set of changes to keep.

When the user uploads a document to Google Cloud Connect, the service inserts some metadata into the file. Metadata is information about other information. In this case, the metadata identifies the file so that changes will track across all copies. The back end is similar to the Google File System and relies on the Google Docs infrastructure. As the documents sync to the master file, Google Cloud Connect sends the updated data out to all downloaded copies of the document using the metadata to guide updates to the right files.

11.3.4 Google Cloud Print

Google Cloud Print is a service that extends the printer's function to any device that can connect to the Internet. To use Google Cloud Print, the user needs to have a free Google profile, an app, a program, or a website that incorporates the Google Cloud Print feature, a cloud-ready printer or printer connected to a computer logged on to the Internet.

When Google Cloud Print is used through an app or website, the print request goes through the Google servers. Google routes the request to the appropriate printer associated with the user's Google account. Assuming the respective printer is on and has an active Internet connection, paper, and ink, the print job should execute on the machine. The printer can be shared with other people for receiving documents through Google Cloud Print.

Because most printers are not cloud ready, most Google Cloud Print users will need to have a computer act as a liaison. Google Cloud Print is an extension built into the Google Chrome Browser, but it should be enabled explicitly. Once enabled, the service activates a small piece of code called a connector. The connector's job is to interface between the printer and the outside world. The connector uses the user's computer printer software to send commands to the printer.

If one has a cloud-ready printer, one can connect the printer to the Internet directly without the need for a dedicated computer. The cloud printer has to be registered with Google Cloud Print to take advantage of its capabilities.

Because Google allows app and website developers to incorporate Google Cloud Print into their products as they see fit, there is no standard approach to executing a print job. Google Cloud Print depends on developers incorporating the feature into their products. Not every app or site will have Google Cloud Print built into it, which limits its functionality. Naturally, Google builds the service into its own products, but many people rely on services from multiple sources and may find Google Cloud Print does not have a wide enough adoption to meet all their needs.

11.3.5 Google App Engine

Google App Engine lets the user run web applications on Google's infrastructure. App Engine applications are easy to build, easy to maintain, and easy to scale as traffic and data storage needs grow. With App Engine, there are no servers to maintain: Just upload the application, and it is ready to serve users.

The app can be served from the user's own domain name (such as http://www.example.com/) using Google Apps. Otherwise, it can be served using a free name on the appspot.com domain. An application can be shared with the world or limit access to members of an organization. Figure 11.2 shows the different modules in Google App Engine. Integration of cloud computing services with support services and client capabilities is shown in the diagram.

Google App Engine supports apps written in several programming languages. With App Engine's Java runtime environment, one can build one's app using standard Java technologies, including the JVM, the Java servlets, and the Java programming language—or any other language. App Engine also features a Python runtime environment, which includes a fast Python interpreter and the Python standard library. App Engine also features a PHP runtime, with native support for Google Cloud SQL and Google Cloud Storage that works just like using a local MySQL instance and doing local file writes. Finally, App Engine

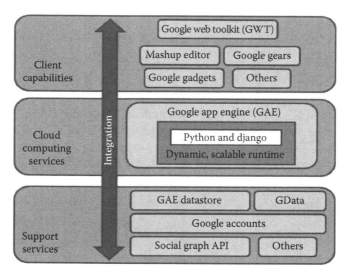

FIGURE 11.2
Google App Engine. (Adapted from http://rdn-consulting.com/blog/tag/azure/, accessed January 16, 2014).

provides a Go runtime environment that runs natively compiled Go code. These runtime environments are built to ensure that your application runs quickly, securely, and without interference from other apps on the system.

With App Engine also, the user has to only pay for what he or she uses. There are no setup costs and no recurring fees. The resources used by the application such as storage and bandwidth are measured in gigabyte and billed at competitive rates. One has to control the maximum amount of resources one's app can consume, so it always stays within one's budget.

App Engine costs nothing to get started. All applications can use up to 1 GB of storage and enough CPU and bandwidth to support an efficient app serving around five million page views a month, absolutely free. When billing is enabled for the application, free limits are raised, and one has to only pay for resources one uses above the free levels.

11.4 Amazon Web Services

Amazon Web Services (AWS) is a collection of remote computing services (also called web services) that together make up a cloud computing platform, offered over the Internet by Amazon.com. The most central and well known of these services are Amazon Elastic Compute Cloud (Amazon EC2), Amazon Simple Queue Service (Amazon SQS), and Amazon S3 as shown in Figure 11.3.

Amazon EC2 is a computing service, whereas Amazon SQS and Amazon S3 are support services. The service is advertised as providing a large

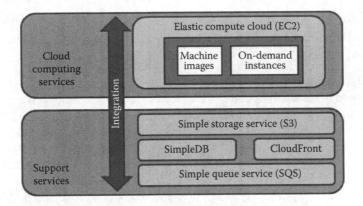

FIGURE 11.3
AWS. (Adapted from http://rdn-consulting.com/blog/tag/azure/, accessed January 16, 2014).

computing capacity (potentially many servers) much faster and cheaper than building a physical server farm. Amazon's data centers are located at Ashburn, Virginia, Dallas/Fort Worth, Los Angeles, Miami, Newark, New Jersey, Palo, Alto, California, Seattle, St. Louis, Amsterdam, Dublin, Frankfurt, London, Hong Kong, Singapore, Tokyo, etc.

11.4.1 Amazon Elastic Compute Cloud

Amazon EC2 is an IaaS offered by AWS and is the leading provider of IaaS in the current market. Powered by a huge infrastructure that the company has built to run its retail business, Amazon EC2 provides a true virtual computing environment. By providing a variety of virtual machine or instance types, operating systems, and software packages to choose from, Amazon EC2 enables the user to instantiate virtual machines of his choice through a web service interface. The user can change the capacity and characteristics of the virtual machine by using the web service interfaces, hence named *elastic*. Computing capacity is provided in the form of virtual machines or server instances by booting Amazon Machine Images (AMI), which can be instantiated by the user. An AMI contains all the necessary information needed to create an instance. The primary Graphical User Interface (GUI) interface is the AWS Management Console (point and click) and a web service API that supports both Simple Object Access Protocol and Query Requests. The API provides programming libraries and resources for Java, PHP, Python, Ruby, Windows, and .Net. The infrastructure is virtualized by using Xen hypervisor, and different instance types are provided as follows:

- Standard instances—suitable for most applications
- Micro instances—suitable for low-throughput applications
- High-memory instances—suitable for high-throughput applications

- High-CPU instances—suitable for compute-intensive applications
- Cluster compute instances—suitable for high-performance computing (HPC) applications

The instances can be obtained on demand on an hourly basis, thus eliminating the need of forecasting computing needs earlier. Instances can be reserved earlier, and a discounted rate is charged for such instances. Users can also bid on unused Amazon EC2 computing capacity and obtain instances. Such instances are called as Spot Instances. Those bids that exceed the current Spot Price is provided with the instance, which allows the user to reduce costs. The Spot Price is varying and is decided by the company.

Instances can be placed in multiple locations, which are defined by regions and availability zones. Availability zones are distinct locations that are engineered to be insulated from failures in other availability zones and provide inexpensive, low-latency network connectivity to other availability zones in the same region. Thus, placing the instances in multiple locations enables fault tolerance and failover reliability. The Amazon EC2 instances can be monitored and controlled by the AWS Management Console and the web service API. However, AWS provides Amazon Cloud Watch, a web service that provides monitoring for AWS cloud resources, starting with Amazon EC2. It provides customers with visibility into resource utilization, operational performance, and overall demand patterns—including metrics such as CPU utilization, disk reads and writes, and network traffic.

Instances are authenticated using a signature-based protocol, which uses key pairs. Another important feature provided is the Amazon Virtual Private Cloud (Amazon VPC). The existing IT infrastructure can be connected to Amazon EC2 via a virtual private network (VPN). Isolated computing resources are provided in Amazon VPC, and the existing management capabilities such as security services, firewalls, and intrusion detection systems can be extended to isolated resources of Amazon EC2.

Elastic load balancing (ELB) enables the user to automatically distribute and balance the incoming application's traffic among the running instances based on metrics such as request count and request latency. Fault tolerance and automatic scaling can be performed by configuring the ELB as per the specific needs. ELB monitors the health of the instances running and routes traffic away from a failing instance.

An instance is stored as long as it is operational and is removed on termination. Persistent storage can be enabled by using either Elastic Block Storage (EBS) or Amazon Simple Storage Service (S3). EBS provides a highly reliable and secure storage, and the storage volumes can be used to boot an Amazon EC2 instance or be attached to an instance as a standard block device. Amazon S3 provides a highly durable storage infrastructure designed for mission-critical and primary data storage. Storage is based on units called objects whose size can vary from one byte to five gigabytes of data. These objects are stored in a bucket and retrieved via a unique, developer-assigned key.

It is accessible through a web service interface and provides authentication procedures to protect against unauthorized access.

11.4.2 Amazon Simple Storage Service

Amazon Simple Storage Service known as Amazon S3, is the storage for the Internet. It is designed to make web-scale computing easier for developers. Amazon S3 provides a simple web service interface that can be used to store and retrieve any amount of data, at any time, from anywhere on the web. It gives any developer access to the same highly scalable, reliable, secure, fast, inexpensive infrastructure that Amazon uses to run its own global network of websites. The service aims to maximize benefits of scale and to pass those benefits on to developers.

Along with its simplicity, it also takes care of other features like security, scalability, reliability, performance, and cost. Thus, Amazon S3 is a highly scalable, reliable, inexpensive, fast, and also easy to use service that meets design requirements and expectations.

Amazon S3 provides a highly durable and available store for a variety of content, ranging from web applications to media files. It allows users to offload storage where one can take advantage of scalability and pay-as-you-go pricing. For sharing content that is either easily reproduced or where one needs to store an original copy elsewhere, Amazon S3's Reduced Redundancy Storage (RRS) feature provides a compelling solution. It also provides a better solution in the case of storage for data analytics. Amazon S3 is an ideal solution for storing pharmaceutical data for analysis, financial data for computation, and images for resizing. Later this content can be sent to Amazon EC2 for computation, resizing, or other large-scale analytics without incurring any data transfer charges for moving the data between the services.

Amazon S3 offers a scalable, secure, and highly durable solution for backup and archiving critical data. For data of significant size, the AWS Import/Export feature can be used to move large amounts of data into and out of AWS with physical storage devices. This is ideal for moving large quantities of data for periodic backups, or quickly retrieving data for disaster recovery scenarios. Another feature offered by Amazon S3 is its Static Website Hosting, which is ideal for websites with static content, including html files, images, videos, and client-side scripts such as JavaScript.

11.4.3 Amazon Simple Queue Service

Another service of AWS is Amazon SQS. It is a fast, reliable, scalable, fully managed message queuing service. SQS makes it simple and cost effective to decouple the components of a cloud application. SQS can be used to transmit any volume of data, at any level of throughput, without losing messages or requiring other services to be always available.

Amazon SQS is a distributed queue system that enables web service applications to quickly and reliably queue messages that one component in the application generates to be consumed by another component. A queue is a temporary repository for messages that are waiting to be processed.

Amazon SQS offers various features like allowing multiple readers and writers at the same time, providing access control facilities, guaranteeing high availability of sending, and retrieving messages due to redundant infrastructure. It also gives provision for having variable length messages as well as configurable settings for each queue.

11.5 Microsoft

Cloud computing provides a new way of looking at IT at Microsoft called Microsoft IT (MSIT). Cloud computing is now the preferred and default environment for new and migrated applications at Microsoft. MSIT has developed a methodology and a set of the best practices for analyzing their current application portfolio for possible candidates to migrate to cloud computing. This analysis enables MSIT to select the ideal cloud computing–based environment for each application. MSIT has captured these best practices and documented them for other Microsoft customers who wish to migrate their organizations to cloud computing.

11.5.1 Windows Azure

Windows Azure Cloud Services (web and worker roles/PaaS) allow developers to easily deploy and manage application services. It delegates the management of underlying role instances and operating system to the Windows Azure platform.

The Migration Assessment Tool (MAT) for Windows Azure encapsulates all the information to be aware of before attempting the application migration to Windows Azure. Based on the response to a series of simple binary questions, the tool generates a report that outlines the amount of development effort involved to migrate the application, or the architecture considerations for a new application.

The Windows Azure Pricing Calculator analyzes an application's potential public cloud requirements against the cost of the application's existing infrastructure. This tool can help to compare current operational costs for an application, against what the operating costs would be on Windows Azure and SQL Azure.

Windows Azure Pack for Windows Server is a collection of Windows Azure technologies available to Microsoft customers at no additional cost for installation into their data center. It runs on top of Windows Server 2012

R2 and System Center 2012 R2 and, through the use of the Windows Azure technologies, it allows you to offer a rich, self-service, multitenant cloud, consistent with the public Windows Azure experience.

11.5.2 Microsoft Assessment and Planning Toolkit

The Microsoft Assessment and Planning Toolkit (MAP) is an agentless, automated, multiproduct planning and assessment tool for cloud migration. MAP provides detailed readiness assessment reports, executive proposals, and hardware and software information. It also provides recommendations to help organizations accelerate the application migration process for both private and public cloud planning assessments. MAP analyzes server utilization data for server virtualization and also server consolidation with Hyper-V.

11.5.3 SharePoint

Microsoft offers its own online collaboration tool called SharePoint. Microsoft SharePoint is a web application platform that comprises a multipurpose set of web technologies backed by a common technical infrastructure. By default, SharePoint has a Microsoft Office–like interface, and it is closely integrated with the Office suite. The web tools are designed to be usable by nontechnical users. SharePoint can be used to provide intranet portals, document and file management, collaboration, social networks, extranets, websites, enterprise search, and business intelligence. It also has system integration, process integration, and workflow automation capabilities. Unlike Google Cloud Connect, Microsoft SharePoint is not a free tool. But it has additional features that cannot be matched by Google or any other companies.

11.6 IBM

IBM is one among the players in the field of cloud computing offering various cloud services to the consumers. IBM cloud computing consists of cloud computing solutions for enterprises as offered by the global IT company IBM. All offerings are designed for business use, marketed under the name IBM SmartCloud. IBM cloud includes IaaS, SaaS, and PaaS offered through public, private, and hybrid cloud delivery models, in addition to the components that make up those clouds.

IBM offers an entry point to cloud computing whether a client is designing their own virtual private cloud, deploying cloud service, or consuming cloud workload applications. The IBM cloud framework begins with the physical hardware of the cloud. IBM offers three hardware platforms for cloud computing, which offer built-in support for virtualization. The next layer of the IBM framework is virtualization. IBM offers IBM Websphere application infrastructure solutions that support programming models and open standards for virtualization.

The management layer of the IBM cloud framework includes IBM Tivoli middleware. Management tools provide capabilities to regulate images with automated provisioning and deprovisioning, monitor operations, and meter usage while tracking costs and allocating billing. The last layer of the framework provides integrated workload tools. Workloads for cloud computing are services or instances of code that can be executed to meet specific business needs. IBM offers tools for cloud-based collaboration, development and test, application development, analytics, business-to-business integration, and security.

11.6.1 Cloud Models

IBM offers a spectrum of cloud delivery options ranging from solely private cloud to solely public cloud and numerous variations in between. IBM gives the option to build a customized cloud solution out of a combination of public cloud and private cloud elements. Companies that prefer to keep all data and processes behind their own firewall can choose a private cloud solution managed by their own IT staff. A company may also choose pay-as-you-go pricing that allows them to run lower-profile applications on a secure public cloud model. Hybrid cloud options allow for some processes to be hosted and managed by IBM, while others are kept on a private cloud or on a VPN or Virtual Local Area Network. IBM also offers planning and consultation throughout the deployment process. Cloud computing is the best choice for mobile software. IBM offers five different cloud provision models:

1. Private cloud, owned and operated by the customer
2. Private cloud, owned by the customer but operated by IBM (or another provider)
3. Private cloud, owned and operated by IBM (or another provider)
4. Virtual private cloud services, based on multitenant support for individual enterprises
5. Public cloud services, based on the provision of functions to individuals

The majority of cloud users choose a hybrid cloud model, with some workloads being served by internal systems, some from commercial cloud providers, and some from public cloud service providers.

For enterprise customers who perceive that the security risk of cloud computing adoption is too high, IBM specializes in secure private cloud offerings.

For building strictly private clouds, IBM offers IBM Workload Deployer and Cloudburst as ready-to-deploy, *cloud in a box*–style solutions. Cloudburst provides blade servers, middleware, and virtualization for an enterprise to build its own cloud-ready virtual machines. Workload Deployer connects an enterprise's existing servers to virtualization components and middleware in order to help deploy standardized virtual machines designed by IBM. For customers who prefer to perform their own integration of private clouds, IBM offers a choice of hardware and software building blocks, along with recommendations and reference architecture, leading the way to deployment. Clients may choose from IBM virtualization–enabled servers, middleware, and SaaS applications.

11.6.2 IBM SmartCloud

IBM SmartCloud is a branded ecosystem of cloud computing products and solutions from IBM. It includes IaaS, SaaS, and PaaS offered through public, private, and hybrid cloud delivery models. IBM places these offerings under three umbrellas: SmartCloud Foundation, SmartCloud Services, and SmartCloud Solutions. Figure 11.4 briefly explains the architecture of IBM SmartCloud.

SmartCloud Foundation consists of the infrastructure, hardware, provisioning, management, integration, and security that serve as the underpinnings of a private or hybrid cloud. Built using those foundational components, PaaS, IaaS, and backup services make up SmartCloud Services. Running on this cloud platform and infrastructure, SmartCloud Solutions consist of a number of collaboration, analytics, and marketing SaaS applications.

Along with IaaS, PaaS, and SaaS, IBM also offers Business Process as a Service (BPaaS). Infrastructure cloud services provide the consumer the provision of processing, storage, networks, and other fundamental computing resources where the consumer is able to deploy and run arbitrary software, which can include operating systems and applications. In platform cloud services, a consumer can deploy consumer-created or consumer-acquired applications onto the cloud infrastructure created using programming languages and tools supported by the provider. Application cloud services allow consumers to use the provider's applications running on a cloud infrastructure. The applications are accessible from various client devices through a thin client interface such as a web browser (e.g., web-based e-mail). Business process cloud services are any business process (horizontal or vertical) delivered through the cloud service model (multitenant, self-service provisioning, elastic scaling, and usage metering or pricing) via the Internet with access via web-centric interfaces and exploiting web-oriented cloud architecture. The BPaaS provider is responsible for the related business functions.

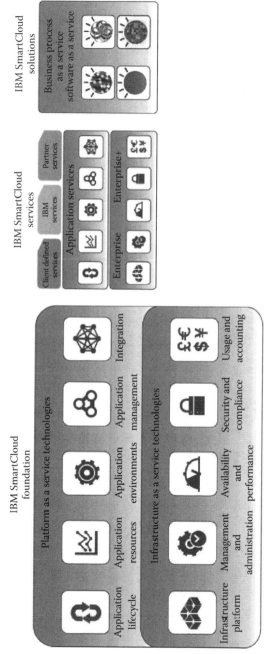

FIGURE 11.4

Architecture of IBM SmartCloud. (Adapted from Transitioning to IBM smart cloud notes, Smart Cloud White Paper-IBM.)

11.7 SAP Labs

SAP Labs makes enterprise software to manage business operations and customer relations. SAP is the leader in the market of enterprise applications in terms of software and software-related service. The company's best-known software products are its enterprise resource planning application systems and management (SAP ERP), its enterprise data warehouse product—SAP Business Warehouse (SAP BW), SAP Business Objects software, and most recently, Sybase mobile products and in-memory computing appliance SAP HANA. SAP is one of the largest software companies in the world.

11.7.1 SAP HANA Cloud Platform

SAP HANA Cloud Platform is an open-standard, Eclipse-based, modular PaaS. In SAP HANA Cloud Platform, applications are deployed via command-line tools to the cloud as web application archive (WAR) files or OSGi bundles. OSGi bundles are normal jar components with extra manifest headers. The applications run within the Java-based SAP HANA Cloud Platform runtime environment. It is powered by SAP HANA and can be maintained using web-based management tools.

The main features of SAP HANA Cloud Platform are as follows:

- Enterprise platform built for developers
- Native integration with SAP and non-SAP software
- In-memory persistence
- Secure data platform
- Lightweight, modular runtime container for applications

SAP HANA Cloud Platform lets the users quickly build and deploy business and consumer applications that deliver critical new functionality to meet emerging business needs. It also helps connect users with customers in more engaging experiences. It provides connectivity based on the cloud connectivity service. As a result, the platform streamlines the integration of new applications at the lowest possible total cost of ownership. Support for open programming standards provides a low barrier entry for developers. This makes them productive from the start in building enterprise applications that can integrate with any SAP or non-SAP solution. No new coding skills are required to work with SAP HANA.

11.7.2 Virtualization Services Provided by SAP

ERP virtualization increases a project's return on investment by maximizing hardware utilization. The business benefits of virtualization of ERP applications are shorter development cycles, reduction in IT costs, improved availability, and energy saving. A joint service from SAP and VMware helps in transition to a more open and flexible private cloud platform based on proven virtualization technology.

11.8 Salesforce

Salesforce.com is a cloud computing and social enterprise SaaS provider based in San Francisco. Of its cloud platforms and applications, the company is best known for its Salesforce CRM product, which is composed of Sales Cloud, Service Cloud, Marketing Cloud, Force.com, Chatter, and Work.com. In addition to its products and platforms, Salesforce.com created AppExchange, a custom application building and sharing platform. The company also has consulting, deployment, and training services.

11.8.1 Sales Cloud

Sales Cloud refers to the *sales* module in Salesforce.com. It includes Leads, Accounts, Contacts, Contracts, Opportunities, Products, Pricebooks, Quotes, and Campaigns (limits apply). It includes features such as web-to-lead to support online lead capture, with autoresponse rules. It is designed to be a start-to-end setup for the entire sales process. Sales Cloud manages contact information and integrates social media and real-time customer collaboration through Chatter. The Sales Cloud gives a platform to connect with customers from complete, up-to-date account information to social insights, all in one place and available anytime, anywhere. Everything is automatically pushed in real time, from contact information to deal updates and discount approvals.

Salesforce.com created the Sales Cloud to be as easy to use as a consumer website like Amazon and built it in the cloud to eliminate the risk and expense associated with traditional software. With its open architecture and automatic updates, the Sales Cloud does away with the hidden costs and drawn-out implementations of traditional CRM software. By continuing to innovate and embrace technologies like mobile, collaboration, and social intelligence, the Sales Cloud has continued to pull ahead of the competition.

11.8.2 Service Cloud: Knowledge as a Service

Service Cloud refers to the *service* (as in *customer service*) module in Salesforce. com. It includes Accounts, Contacts, Cases, and Solutions. It also encompasses features such as the public knowledge base, web-to-case, call center, and self-service portal, as well as customer service automation. Service Cloud includes a call center–like case tracking feature and a social networking plug-in for conversation and analytics.

The Service Cloud delivers the world's first enterprise-grade knowledge base to run entirely on an advanced, multitenant cloud platform. That means one can get all the cloud computing benefits that Salesforce.com is known for delivering without expensive data centers or software. Just powerful knowledge management, without the hassle of on-premises software, is provided. Unlike stand-alone applications, this knowledge base is fully integrated with everything else. Service Cloud has to offer all the tools one needs to run the entire service operation. When the consumer's knowledge base is a core part of CRM solution, knowledge as a process can be managed. One can continually create, review, deliver, analyze, and improve the knowledge. And, because it is delivered by the Service Cloud, user's knowledge is available wherever other customers need it. Agents have the right answers at their fingertips to communicate over the phone, send out through an e-mail, or share via a chat client. The same knowledge base serves up answers to the service website is a part of company's public site. If one wants to take advantage of social channels like Twitter or Facebook, one can easily share knowledge that is tapped into the wisdom of the crowd to capture new ideas or answers. All this is done securely.

The Service Cloud gives the tools that are needed to manage knowledge at enterprise scale. But it also delivers the same great ease of use that Salesforce.com is known for. That means user will benefit no matter what size or how complex the business is.

11.9 Rackspace

Rackspace Cloud, a part of Rackspace, is another player in the cloud computing market. Offering IaaS to clients, it has been used by a large number of enterprises. Rackspace Cloud offers three cloud computing solutions—Cloud Servers, Cloud Files, and Cloud Sites. Cloud Servers provide computational power on demand in minutes; Cloud Sites are for robust and scalable web hosting, and Cloud Files are for elastic online file storage and content delivery.

Cloud Servers is an implementation of IaaS where the computing capacity is provided as virtual machines that run in the Cloud Servers systems. The virtual machine instances are configured with different amounts of capacities. The instances come in different flavors and images. A flavor is an available hardware configuration for a server. Each flavor has a unique combination of disk space, memory capacity, and priority for CPU time. A varied set of instances are available for the user to choose from.

These virtual machines are instantiated using images. An image is a collection of files used to create or rebuild a server. A variety of prebuilt operating system images are provided by Rackspace Cloud (64-bit Linux distributions—Ubuntu, Debian, Gentoo, CentOS, Fedora, Arch, and Red Hat Enterprise Linux) or Windows Images (Windows Server 2008 and Windows Server 2003). These images can be customized to the user's choice to create custom images.

The Cloud Servers systems are virtualized using the Xen Hypervisor for Linux and Xen Server for Windows. The virtual machines that are generated come in different sizes and measured based on the amount of physical memory reserved. Currently, the physical memory can vary from 256 MB to 15.5 GB. In the event of availability of extra CPU power, Rackspace Cloud claims to provide extra processing power to the running workloads, free of cost.

Backup schedules can be created to define when to create server images. This is a useful feature, which enables the user to continue work in the event of failures by using the backup images. Custom images are helpful in creating backup schedules. A type of images, referred to as *gold* server images, can be produced if the servers of that configuration are to be instantiated frequently.

Cloud Servers can be run through the Rackspace Cloud Control Panel (GUI) or programmatically via the Cloud Server API using a RESTful interface. The control panel provides billing and reporting functions and provides access to support materials including developer resources, a knowledge base, forums, and live chat. The Cloud Servers API was open sourced under the Creative Commons Attribution 3.0 license. Language bindings via high-level languages like C++, Java, Python, or Ruby that adhere to the Rackspace specification will be considered as Rackspace-approved bindings. The virtual machine instances are authenticated in the API by a token-based protocol that uses the HTTP x-Header. Private/public keys are used to ensure Secured Shell Access.

Cloud Servers scale automatically to balance load. This process is automated and initiated from either the Rackspace Cloud Control Panel or the Cloud Server API. The amount to scale is specified; the Cloud Server is momentarily taken offline; the RAM, disk space, and CPU allotment are adjusted; and the server is restarted. A Cloud Server can be made to act as a load balancer using simple readily available packages from any of the distribution repositories. Rackspace Cloud is working on beta version of the Cloud Load Balancing product, which provides a complete load balancing solution.

Cloud Servers are provided persistent storage through RAID10 disk storage; thus, data persistency is enabled leading to better functioning.

11.10 VMware

VMware, a leader in virtualization technology, has come up with enterprise cloud computing solutions. Having been a dominating player in the virtualization domain, VMware is currently providing a range of products for the development of private and public clouds and for leveraging the services offered by both as a hybrid cloud, such as VMware vCloud Director, VMware vCloud Datacenter Services, VMware vSphere, and VMware vShield to name a few.

Private clouds enable the better usage and management of internal IT infrastructure than the traditional methods. Greater operational efficiency, secure, fault-tolerant, well-managed computing environments can be modeled and operated. VMware's private cloud offering provides greater standardization, rapid provisioning, and self-service for all applications and unparalleled cost savings by consolidating their physical infrastructures. VMware's modular technology enables the user to select from a variety of hardware, software, and certified service providers to result in efficient cloud computing. Thus, the family of products offered by VMware promotes compatibility and retains the choice of freedom for the users to obtain desired services.

Private clouds can be created by using the VMware vSphere and VMware vCloud Director. VMware vSphere is a robust virtualization platform used to transform IT infrastructures into virtual storage, compute, and network resources and provide them as a service within the organization. VMware vSphere provides services at both the infrastructure and application levels. At the infrastructure level, it provides options to perform efficient operation and management of the compute, storage, and network resources. At the application level, service-level controls are provided for the applications running on the underlying infrastructures, leading to available, secure, and scalable applications.

The VMware vCloud Director, coupled with VMware vSphere, is a software solution that enables enterprises to build secure, multitenant private clouds by pooling infrastructure resources into virtual datacenters and exposing them to users through web-based portals and programmatic interfaces as fully automated, catalog-based services. VMware vCloud Director abstracts the virtual computing environment from the underlying resources and provides a multitenant architecture that features isolated virtual resources, independent LDAP authentication, specific policy controls, and unique catalogs. VMware vShield technologies are used to provide security to these environments by using services like perimeter protection, port-level firewall, NAT and DHCP services, site-to-site VPN, network isolation, and web load balancing. The VMware vCloud Director allows users to catalog infrastructure and application services of the desired configurations and deploy and consume them as needed. Interactions with the virtual data centers or the catalogs are through a user-friendly web portal or the vCloud API. The vCloud API is an open, REST-based API that provides scripted access, complying with the open virtualization format (OVF). The API can be used along with VMware vCenter Orchestrator to automate and orchestrate operational processes like routine tasks, activities, and workflows.

Public and hybrid cloud solutions are provided by VMware by partnering with other companies, certified as service providers. VMware vCloud Datacenter Services and VMware vCloud Express offer efficient solutions for utilizing IaaS either as a public cloud or a hybrid cloud. vCloud Datacenter Services provides a scalable environment, where internal resources are augmented with the external resources. vCloud Datacenter Services are built on the same technology and foundations as VMware vCloud Director and VMware vSphere to enable interoperability between cloud environments. Thus, the user is free to burst his private cloud into public cloud of his preferred service provider.

vCloud Express is an IaaS offering delivered by leading VMware service provider partners. It is a cobranded service that provides reliable, on-demand, pay-as-you-go infrastructure. The VMware vCloud Express providers are Virtacore vCloud Express, Hosting.com, Melbourne IT, and Terremark's vCloud Express. Instance types, load balancing, storage options, and pricing vary between service providers.

11.11 Manjrasoft

Manjrasoft is one of the nonmajor providers of cloud services. But it has come up with a platform called *Aneka* that provides a set of services that help the development of applications in an easier way. Manjrasoft develops market-oriented cloud computing platforms that allow one to build, accelerate, and manage the applications ultimately saving one's time and money, leading to enhanced business productivity and profit.

11.11.1 Aneka Platform

Aneka provides a set of services that make enterprise cloud construction and development of applications as easy as possible without sacrificing flexibility, scalability, reliability, and extensibility.

Figure 11.5 gives an overview of the Aneka platform. The key features supported by *Aneka* are as follows:

1. A configurable and flexible execution platform (container) enabling pluggable services and security implementations. Multiple authentication/authorization mechanisms such as role-based security and Windows domain–based authentication are considered for this purpose.
2. Multiple persistence options including Relational Database Management System (RDBMS), Structured Query Language (SQL) Express, MySQL, and flat files.
3. Software development kit (SDK) supporting multiple programming models including object-oriented thread model, task model for legacy applications, and MapReduce model for data-intensive applications.
4. Custom tools such as Design Explorer for parameter sweep studies.

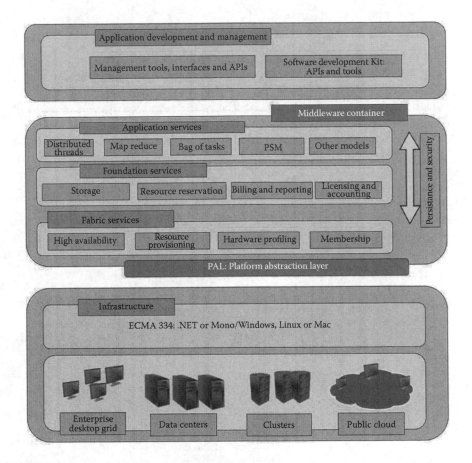

FIGURE 11.5
Overview of the Aneka platform. (Adapted from http://www.manjrasoft.com/aneka_architecture.html, accessed January 8, 2013.)

5. Easy to use management tool for SLA and Quality of Service (QoS) negotiation and dynamic resource allocation.

6. Supports deployment of applications on private or public clouds in addition to their seamless integration.

Aneka allows servers and desktop PCs to be linked together to form a very powerful computing infrastructure. This allows companies to become energy efficient and save money without investing in a number of computers to run their complex applications.

Each *Aneka* node consists of a configurable container that includes information and indexing, scheduling, execution, and storage services. *Aneka* supports multiple programming models, security, persistence, and communications protocols.

11.12 Summary

In this chapter, we have discussed about various companies that support cloud computing by providing tools and technologies to adapt to the cloud environment. Each section briefly describes the cloud features supported in these companies. Few of the services like Google Docs and Google Cloud Print are free, whereas that of AWS, Microsoft, etc., are proprietary. Based on the specific requirements, the user has to make a trade-off between open source and closed source tools/services. An attempt has been made to list the tools/services offered by each company in Table 11.1.

Though there are a number of companies, we have chosen few companies that have progressed a lot in this area. Table 11.2 gives the information about few providers and the prices on a per hour basis along with the service

TABLE 11.1

Tools and Services Offered by Companies

Company Name	Tools/Services
EMC	Captiva Cloud toolkit
Google	Google App Engine, Google Docs, Google Cloud Connect Google Cloud Print
Amazon	Amazon EC2, Amazon S3, Amazon SQS
Microsoft	Microsoft Assessment and Planning Toolkit, Windows Azure Sharepoint
IBM	IBM Smart Cloud
Salesforce	Sales Cloud, Service Cloud
SAP LABS	SAP HANA Cloud
VMware	vCloud
Manjrasoft	Aneka Platform
Red Hat	OpenShift Enterprise, OpenShift Origin
Gigaspaces	Cloudify

TABLE 11.2

Details of Cloud Service Providers

Provider Name	Service Model	Deployment Model	Server Operating System
Amazon Web Services	IaaS	Public	Widows, Linux
Google App Engine	PaaS	Public	Windows
Windows Azure	IaaS	Public	Widows, Linux
IBM Cloud	IaaS	Private, hybrid	Widows, Linux
Salesforce Platform	PaaS	Public	Widows, Linux
Rackspace	IaaS	Public, private, hybrid	Widows, Linux
SAP HANA Cloud	PaaS	Public	Linux

model and deployment model offered by that system. Further details regarding the companies can be seen in the references section.

Review Points

- *MDW*: It is an entry point in the Captiva Cloud toolkit. This module helps to import documents from specified folders/repository (see Section 11.2.2).
- *REST*: It is an architectural style in a distributed system. Google Cloud Storage is a RESTful online storage web service (see Section 11.3.2).
- *AMI*: It is a special type of virtual appliance that is used to create a virtual machine in Amazon EC2 (see Section 11.4.1).
- *RRS*: It is a new storage solution by Amazon S3 that enables cost-effective storage of noncritical data by reducing the redundancy. This feature provides a better solution in the case of storage for data analytics in Amazon SQS (see Section 11.4.2).
- *MAT*: It is a tool provided by Microsoft Azure that addresses migration considerations including app server, database, integration, security, and instrumentation for different platforms (see Section 11.5.1).

Review Questions

1. What do you mean by cloud service provider? Which are the major cloud service providers?
2. List the tools/services provided by Microsoft and explain them in brief.
3. What is Google Cloud Print? What are its advantages?
4. Explain SAP HANA Cloud in brief.
5. What are the services offered by EMC IT? Explain.
6. Explain the services provided by IBM SmartCloud.
7. What are the support services offered by Amazon Web Services? Explain.
8. What do you mean by *Knowledge as a Service*? Which company provides this service? Explain.
9. Explain the features of Aneka.
10. What is vCloud? Explain in brief.

References

1. EMC IT's journey to the private cloud, applications and the cloud experience. White Paper-EMC.
2. http://rdn-consulting.com/blog/tag/azure/. Accessed January 16, 2014.
3. Transitioning to IBM SmartCloud notes. SmartCloud White Paper-IBM.
4. http://www.manjrasoft.com/aneka_architecture.html. Accessed January 8, 2014.

Further Reading

Akamai solutions for cloud computing. White Paper-Akamai.
Cloud-based application evaluation, planning, and migration. Article, June 2012.
Dell vCloud featuring trend micro SecureCloud. White Paper-Dell.
EMC IT's journey to the cloud: A practitioner's guide. White Paper-EMC.
Getting cloud computing right. IBM Global Technology Services, White Paper.
http://googcloudlabs.appspot.com/.
Service Cloud. http://www.salesforce.com/in/service-cloud/overview/.
Cloud Storage. https://cloud.google.com/products/cloud-storage/.
The Service Cloud—Knowledge as a service. White paper-Salesforce.com.
Varia, J. and S. Mathew. Overview of Amazon web services, 2012.
Vecchiola, C., X. Chu, and R. Buyya. Aneka: A software platform for .NET-based cloud computing. In Gentzsch, L., L. Grandinetti, and G. Joubert (eds.), *High Speed and Large Scale Scientific Computing*. IOS Press, Amsterdam, the Netherlands, 2009, pp. 267–295.
VMware vCloud hybrid services. White Paper-VMware.

12

Open Source Support for Cloud

Learning Objectives

The main objective of this chapter is to provide an overview of open source support for cloud computing. After reading this chapter, you will

- Understand the difference between open source and closed source tools
- Know the advantages of open source tools over closed source
- Understand the different open source tools for Infrastructure as a Service (IaaS), Platform as a Service (PaaS), Software as a Service (SaaS), etc.
- Know the available architecture of tools
- Understand the features supported by tools

Preamble

This chapter provides an overview of different open source tools supporting cloud computing. We begin with an introduction to open source tools and their advantages over closed source tools. Subsequent sections provide a brief overview of tools provided for IaaS, PaaS, and SaaS, simulators for research, and distributed computing for managing distributed systems. Subsections in each of these sections explain the architecture and features of the tools. We focus on giving a brief idea about the different tools available as open source. After the end of this chapter, the reader can gain knowledge about the various tools and make appropriate decisions for selecting them.

12.1 Introduction

Cloud computing is one of the most popular technologies nowadays. According to the National Institute of Standards and Technology (NIST), cloud computing is an on-demand, rapid-service provisioning model over the Internet in the form of compute, network, storage, and application with minimal management effort. The characteristics of cloud computing are on-demand self-service, broad network access, resource pooling, rapid elasticity, and measured services. In order to support cloud computing, various tools are freely available and there are proprietary softwares. Open source support for cloud computing is a major developmental step for the rapid growth of cloud technology.

12.1.1 Open Source in Cloud Computing: An Overview

Open source refers to a program or software in which the source code (the form of the program when a programmer writes a program in a particular programming language) is available to the general public for use and/or modification from its original design is free of charge. Open source code is typically created as a collaborative effort in which programmers improve upon the code and share the changes within the community. Open source is free whereas proprietary software is privately owned and controlled. In the computer industry, proprietary is considered as the opposite of open. A proprietary design or technique is one that is owned by a company. It also implies that the company has not divulged specifications that would allow other companies to duplicate the product.

In cloud computing environment, open source support has led to many innovations by providing various things as services, that is, X as a Service, where X can be Software, Platform, Infrastructure, etc.

12.1.2 Difference between Open Source and Closed Source

Some software has a source code that cannot be modified by anyone but the person, team, or organization that created it and maintains exclusive control over it. This kind of software is frequently called *proprietary software* or *closed source* software, because its source code is the property of its original authors, who are the only ones legally allowed to copy or modify it. Microsoft Word and Adobe Photoshop are examples of proprietary software. In order to use proprietary software, computer users must agree (usually by signing a license displayed the first time they run this software) that they will not do anything with the software that the software's authors have not permitted explicitly.

Open source software is different. Its authors make its source code available to others who would like to view that code, copy it, learn from it, alter it, or share it. LibreOffice and the GNU Image Manipulation Program are examples of open source software. As they do with proprietary software, users must accept the terms of a license when they use open source software, but the legal terms of open source licenses differ dramatically from those of proprietary licenses. Open source software licenses promote collaboration and sharing because they allow others to make modifications to source code and incorporate that code into their own projects. Some open source licenses ensure that anyone who alters and then shares a program with others must also share that program's source code without charging a licensing fee for it. In other words, computer programmers can access, view, and modify open source software whenever they like as long as they let others do the same when they share their work. In fact, they could be violating the terms of some open source licenses if they do not do this.

12.1.3 Advantages of Having an Open Source

1. *Larger developer support*: Open source is helpful to develop the platform. It gives much larger support for developers and gives them a feeling of ownership as they can alter whatever they like.

2. *Customizable*: In a closed source scenario, developers are given only options to change what the original developer chooses, whereas open source lets the developers have full control and customize the look feel.

3. *More secure*: Open source is much more transparent than closed source. Anyone can look over the code. Having thousands of people reading through one's code, bugs and vulnerabilities are located much quicker and submitted for fixing. It also lets the user know if the bug has been fixed as one can check the code after each release.

4. *Extended community support*: As a product ages, the original developer might move on and stop developing, leaving the product to age with no new fixes or features, but if it is open, then usually the community takes over and continues working on it allowing the usable life of the product to be extended well beyond what the original developer intended.

12.2 Open Source Tools for IaaS

IaaS is a provision model in which an organization outsources the equipment used to support operations, including storage, hardware, servers, and networking components. The service provider owns the equipment and is responsible for housing, running, and maintaining it. The client typically pays on a per-use basis. The following subsections explain some of the open source tools available for providing IaaS in cloud computing.

12.2.1 OpenNebula

OpenNebula is a flexible tool that orchestrates storage, network, and virtualization technologies to enable the dynamic placement of services on distributed infrastructures. OpenNebula provides a modular architecture intended to be flexible. One of these modules is the *scheduler,* an interesting algorithm that places virtual machines (VMs) depending on their requirements. The client communication is also managed by modules that offer interfaces based on web services. It is perhaps the only open management platform that has invested into a tailorable VM placement algorithm. As such, it may provide a nice environment for those researchers seeking to compare and develop different resource allocation strategies. A limitation found with the OpenNebula is that, like XCP, their infrastructure assumes a classical cluster-like architecture with a front-end and without any redundant services. Figure 12.1 shows the infrastructure virtualization of OpenNebula.

OpenNebula is an open source management tool that helps virtualized data centers oversee private clouds, public clouds, and hybrid clouds. It combines existing virtualization technologies with advanced features for multitenancy, automated provisioning, and elasticity. A built-in virtual network manager maps virtual networks to physical networks. OpenNebula is vendor neutral, as well as platform agnostic and application programming interface (API) agnostic. It can use kernel-based virtual machine (KVM), Xen, or VMware hypervisor.

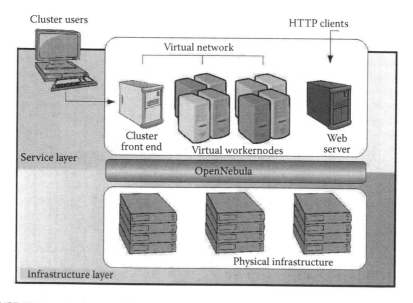

FIGURE 12.1
Infrastructure virtualization in OpenNebula. (Adapted from http://blog.dsa-research. org/?author=7&paged=2.)

Some reasons why OpenNebula is mainly used are listed in the following:

1. It is the most advanced and innovative enterprise-class functionality for the management of virtualized data centers to build private, public, and hybrid clouds.
2. OpenNebula is fully platform independent with broad support for commodity and enterprise-grade hypervisor, storage, and networking resources, allowing to leverage existing IT infrastructure, protecting the investments, and avoiding vendor lock-in.
3. It provides open, adaptable, and extensible architecture, interfaces, and components to build customized cloud service or product.
4. It supports cloud interoperability and portability providing cloud consumers with choice across standards and most popular cloud interfaces.
5. OpenNebula is not a feature- or performance-limited edition of an enterprise version, it is truly an open source code distributed under Apache license.

12.2.2 Eucalyptus

Eucalyptus implements IaaS-style private and hybrid clouds. The platform provides a single interface that lets users access computing infrastructure resources (machines, network, and storage) available in private clouds—implemented by Eucalyptus inside an organization's existing data center—and resources available externally in public cloud services. The software is designed with a modular and extensible web services–based architecture that enables Eucalyptus to export a variety of APIs toward users via client tools. Currently, Eucalyptus implements the industry-standard Amazon Web Services (AWS) API, which allows the interoperability of Eucalyptus with existing AWS services and tools. Eucalyptus provides its own set of command line tools called Euca2ools, which can be used internally to interact with Eucalyptus private cloud installations or externally to interact with public cloud offerings, including Amazon EC2. Figure 12.2 represents the architecture of Eucalyptus Cloud. The main components in this are cluster controller, cloud controller, node controller, walrus storage controller, and storage controller:

1. *Cluster controller* (CC): It manages one or more node controllers and is responsible for deploying and managing instances on them. It communicates with node controller and cloud controller simultaneously. CC also manages the networking for running the instances under certain types of networking modes available in Eucalyptus.
2. *Cloud controller* (CLC): It is the front end for the entire ecosystem. CLC provides an Amazon EC2/S3 compliant web services interface

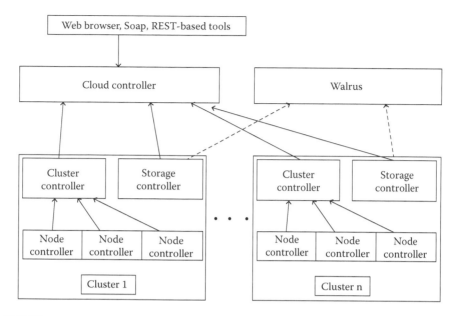

FIGURE 12.2

Eucalyptus Cloud architecture. (Adapted from http://mdshaonimran.wordpress.com/2011/11/26/eucalyptus-and-its-components/.)

to the client tools on one side and interacts with the rest of the components of the Eucalyptus infrastructure on the other side.

3. *Node controller* (NC): It is the basic component for nodes. It maintains the life cycle of the instances running on each nodes. It interacts with the OS, hypervisor, and the CC simultaneously.

4. *Walrus storage controller* (WS3): It is a simple file storage system. WS3 stores the machine images and snapshots. It also stores and serves files using S3 APIs.

5. *Storage controller* (SC): It allows the creation of snapshots of volumes. It provides persistent block storage over AoE or iSCSI to the instances.

While working over Xen and KVM hypervisors, Eucalyptus provides an open source solution to manage the virtual infrastructure of a cloud. The use of interfaces based on web services is one of their key characteristics, allowing native integration with Amazon services. Moreover, the hierarchical architecture is designed to reduce human intervention.

12.2.3 OpenStack

OpenStack, a cloud-computing project, aims to provide IaaS. It is a global collaboration of developers and cloud computing technologists producing the ubiquitous open source cloud computing platform for building public and private clouds. It delivers solutions for all types of clouds by being simple to implement, massively scalable, and feature rich. The technology consists of a series of interrelated projects delivering various components for a cloud infrastructure solution.

The goals of the OpenStack initiative are to support interoperability between cloud services and allow businesses to build Amazon-like cloud services in their own data centers. OpenStack, which is freely available under the Apache 2.0 license, is often referred to in the media as *the Linux of the Cloud* and is compared to Eucalyptus and the Apache CloudStack projects.

Figure 12.3 shows the architecture and project relationships of OpenStack. There are seven core components of OpenStack: compute, object storage, identity, dashboard, block storage, network, and image service.

1. *Object storage* allows user to store or retrieve files. The code name given for this is *swift*. The swift architecture is very distributed to prevent any single point of failure as well as to scale horizontally.

2. *Image* provides a catalog and repository for virtual disk images. These disk images are mostly commonly used in OpenStack Compute. The code name given for this is *glance*. Glance serves a central role to the overall IaaS picture. It accepts API requests for images (or image metadata) from end users or nova components and can store its disk files in the object storage service, swift.

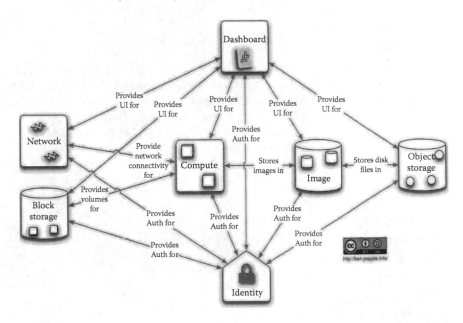

FIGURE 12.3
OpenStack conceptual architecture. (Adapted from http://ken.pepple.info/openstack/2012/09/25/openstack-folsom-architecture/.)

3. *Compute* provides virtual servers upon demand. The code name given for this is *nova*. Nova is the most complicated and distributed component of OpenStack. A large number of processes cooperate to turn end-user API requests into running VMs.

4. *Dashboard* provides a modular web-based user interface (UI) for all the OpenStack services. The code name given for this is *horizon*.

5. *Identity* provides authentication and authorization for all the OpenStack services. It also provides a service catalog of services within a particular OpenStack cloud. The code name given for this is *keystone*. Keystone provides a single point of integration for OpenStack policy, catalog, token, and authentication.

6. *Network* provides a *network connectivity as a service* between interface devices managed by other OpenStack services (most likely nova). The service works by allowing users to create their own networks and then attach interfaces to them. The code name given for this is *neutron*. Neutron interacts mainly with nova, where it provides networks and connectivity for its instances.

7. *Block storage* provides persistent block storage to guest VMs. The code name given for this is *cinder*. Cinder separates out the persistent block storage functionality that was previously part of OpenStack compute (in the form of nova volume) into its own service.

12.2.4 Apache CloudStack

Apache CloudStack is an open source software designed to deploy and manage large networks of VMs, as a highly available, highly scalable IaaS cloud computing platform. CloudStack is used by a number of service providers to offer public cloud services and by many companies to provide an on-premises (private) cloud offering, or as part of a hybrid cloud solution. CloudStack is a better solution that includes almost all the features that most organizations expect from an IaaS cloud. It can be listed as follows:

- Compute orchestration
- Network as a Service
- User and account management
- Full and open native API
- Resource accounting
- UI

12.2.5 Nimbus

Nimbus is a very good open source for IaaS work in administration of a virtual network. It is supported by Secure Shell (SSH) into all compute nodes.

Cloud computing performances depend on different parameters such as the CPU speed, the amount of memory, network, and hard drive speed. In a virtual environment, the hardware is shared between VMs. The features of Nimbus are as follows:

- Data center operators can easily build cloud services within their existing infrastructure to offer on-demand, elastic cloud services.
- It is an open source IaaS software platform that enables users to build, manage, and deploy compute cloud environments.

12.2.6 GoGrid Cloud

GoGrid Cloud is a cloud hosting service that enables automated provisioning of virtual and hardware infrastructures over the Internet. It is an IaaS provider offering virtual and physical servers, storage, networking, load balancing, and security in real time and in multiple data centers. These services are accessed and operated using standard network protocols and IP addresses over the Internet. GoGrid Cloud also provides hybrid hosting, where virtual and physical servers can be provisioned on the same network. Virtual servers can be created from a variety of server images and server sizes, based on RAM allocations. These images come in different sizes—from 0.5 to 16 Gb of RAM. GoGrid Cloud provides standard Windows and Linux operating systems as images, with root/administrator access. Custom images can also be created using the GoGrid MyGSI, an easy and semiautomated process for creating, editing, saving, and deploying a GoGrid server image (GSI). GoGrid has also partnered with various companies, which has created a full suite of server images created and maintained by our partner community. These images come bundled with popular software and applications, providing solutions like disaster recovery, backup solutions, and cloud server security. Similar to Amazon EC2, GoGrid also provides the ability of placing virtual servers in multiple geographic locations or multiple data centers, making failover or disaster recovery possible.

Interaction with the GoGrid infrastructure is through GoGrid portal or GoGrid API. GoGrid portal is a web-based graphical user interface (GUI) that allows users to provision or order cloud (virtual) and dedicated (physical) servers, F5 hardware load balancing, cloud storage, content delivery networks, and custom server images in real time. The GoGrid API is a representational state transfer (REST)–like query interface. Communication is based on the HTTP request methods like GET or POST. The open source API supports Java, PHP, Python, Ruby, C#, and even shell scripting languages such as bash. Cloud management can be performed by using partner GSIs to ensure *full control of the cloud*. Scalability, security, portability, management, and control are core services provided by the partner GSIs.

Load balancing to handle application traffic spike is enabled by fully integrated and redundant f5 load balancers, which can be provisioned from the web portal or API. Application downtime is prevented by spreading traffic among multiple servers. Failover reliability is provided by the f5 load balancers by redirecting existing traffic to online servers in the event of unavailability of a particular server. These f5 load balancers enable rapid scaling to handle drastic increase in traffic and maintain application traffic.

Storage is provided through GoGrid Cloud Storage, an instantly scalable and reliable file-level backup service for Windows and Linux cloud servers running in the cloud. These servers mount the storage volumes via a secure private network and use common protocols to transfer data. GoGrid Cloud Storage supports SCP, FTP, SAMBA/CIFS, and RSYNC. GoGrid Cloud provides security solutions in the form of hardware firewalls. Virtual private network (VPN) tunnels are used to provide secure administrative access to cloud servers, thus protecting them from externals threats.

12.3 Open Source Tools for PaaS

PaaS is a category of cloud computing services that provides a computing platform for development of applications. PaaS offerings facilitate the deployment of applications without the cost and complexity of buying and managing the underlying hardware and software and provisioning hosting capabilities. It may also include facilities for application design, development, testing, and deployment. It also offers services such as web service integration, security, database integration, and storage. The following subsections give an overview of some of the open source tools that provide PaaS.

12.3.1 Paasmaker

Paasmaker is an open source PaaS. It provides full visibility to the users. Actions are broken down into trees that can be monitored all the time. It supports portability by offering interfaces for several common languages like PHP, Ruby, Python, and Node.js. Local development is made as one can run apps from a local directory. Plug-in-based architecture is very much suitable for extending the plug-ins. Paasmaker allows customization via plug-ins to easily extend the system in certain manners to offer additional services or runtimes. It is designed for the clusters. Paasmaker distributes work to multiple machines and monitors them. If a deployment fails, it is routed around automatically. If a controller node fails, the rest of the cluster stays up till it comes back.

12.3.2 Red Hat OpenShift Origin

OpenShift Origin is the open source upstream of OpenShift, the next-generation application hosting platform developed by Red Hat. This is also known as Red Hat's PaaS, OpenShift takes care of infrastructure, middleware, and management. OpenShift Origin includes support for a wide variety of language runtimes and data layers including Java EE6, Ruby, PHP, Python, Perl, MongoDB, MySQL, and PostgreSQL.

OpenShift Origin platform has two basic function units: broker and node servers. Communication between these units is through message queuing service. The broker is the single point of contact for all application management activities. It is responsible for managing user logins, dynamic shutdown (DNS), application state, and general orchestration of the applications. Customers do not contact the broker directly; instead, they use the web console, command line interface (CLI) tools, or the JBoss Tools integrated development environment (IDE) to interact with the broker over a REST-based API. The node servers host the built-in *cartridges* that will be made available to users and the *gears* where user applications will actually be stored and served. MCollective client running on each node is responsible for receiving and performing the actions requested by the broker. OpenShift Origin supports several *built-in* cartridges based on the most popular app development languages and databases. In order for these to work, the underlying technology must be installed on every node server in an Origin system. A gear represents the slice of the node's CPU, RAM, and base storage that is made available to each application. OpenShift Origin supports multiple gear configurations, enabling users to choose from the various gear sizes at application setup time. When an application is created, the broker instructs a node server to create a new gear to contain it. Whenever a new gear is created on a node server, CPU and RAM *shares* are allocated for it and a directory structure is created.

12.3.3 Xen Cloud Platform

The Xen Cloud Platform (XCP) manages storage, VMs, and the network in a cloud. XCP does not provide the overall cloud architecture but rather focuses on configuration and maintenance of clouds. It also enables external tools, including Eucalyptus and OpenNebula, to better leverage the Xen hypervisor. XCP is an open source infrastructure manager tool for clouds that does not provide the overall architecture for cloud computing since it does not provide interfaces to end users to interact with the cloud. However, XCP provides a useful environment for administrators and an API for developers of cloud management systems.

12.3.4 Cloudify

Cloudify is an open source private PaaS from GigaSpaces Technology, Inc. that allows user to deploy, manage, and scale the application. It is a software runtime lifecycle management system for cloud-hosted services. It manages in an automated fashion:

- The provisioning of cloud infrastructure
- The installation and configuration of services on that infrastructure

- Monitors those services for health and parameters. Respond to service-level agreement (SLA) violations
- The uninstallation of service software and deprovisioning of cloud infrastructure
- Provide customized ways to interact with the running system that hide its cloudy deployment

Cloudify is designed to bring any app to any cloud-enabling enterprises, independent software vendors (ISVs), and managed service providers alike to quickly benefit from the cloud automation and elasticity organizations today need. Cloudify helps user to maximize application onboarding and automation by externally orchestrating the application deployment and runtime. Cloudify's DevOps approach treats infrastructure as code, enabling the user to describe deployment and postdeployment steps for any application through an external blueprint (aka, a recipe, which can then be taken from cloud to cloud, unchanged), that is, Cloudify supports portability. Cloudify provides web UI, CLI, and REST API for faster deploying, managing, and configuring the application.

12.4 Open Source Tools for SaaS

SaaS is a software distribution model in which applications are hosted by a vendor or service provider and made available to customers over a network, typically the Internet. SaaS is becoming an increasingly prevalent delivery model as underlying technologies that support web services and mature service-oriented architecture (SOA) and new developmental approaches, such as Ajax, become popular. The following subsections briefly explain some of the open source tools for SaaS.

12.4.1 Apache VCL

VCL stands for virtual computing lab. Apache VCL is an open source solution for the remote access over the Internet to dynamically provision and reserve computational resources for diverse applications, acting as SaaS solution.

VCL has a simple architecture with four main components: web portal, database, management node, and compute node. Figure 12.4 shows the architecture of VCL.

Web server represents the VCL portal and uses Linux/Apache/PHP solution. This portal provides a UI that enable the requesting and management of VCL resources.

Database server stores information about VCL reservations, access controls, and machine and environment inventory. It uses Linux/SQL solution.

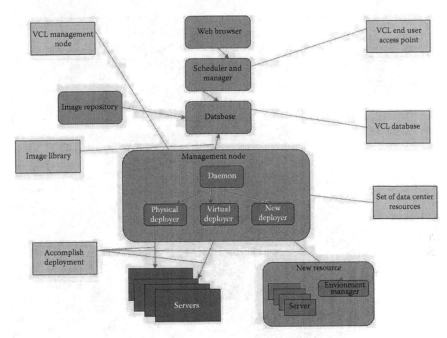

FIGURE 12.4
Architecture of VCL. (Adapted from http://vcl.apache.org/info/architecture.html.)

Management node is the processing engine. A management node controls a subset of VCL resources, which may be physical blade servers, traditional rack, or VMs. It uses Linux/VCLD (perl)/image library solution. VCLD is a middleware responsible to process reservations or jobs assigned by the VCL web portal. According to the type of environment requested, VCLD should assure that service (computational environment) will be available to user.

Compute nodes include physical servers, VMs, computing lab machines as well as cloud compute resource.

Conceptual overview of Apache VCL is given in Figure 12.5. Remote users connect to the VCL Scheduling Application (the web VCL portal) and request access to a desired application environment. The application environment consists of an operating system and a suite of applications. The computer types are machine room blade servers, VMware VMs, and stand-alone machines.

12.4.2 Google Drive

Google Drive is a file storage and synchronization service provided by Google that enables user cloud storage, file sharing, and collaborative editing. Files shared publicly on Google Drive can be searched with web search engines.

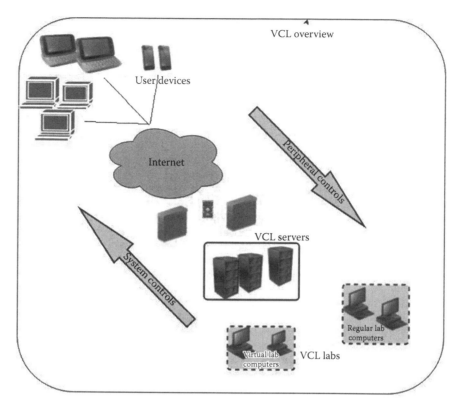

FIGURE 12.5
Apache VCL conceptual overview. (Adapted from https://cwiki.apache.org/confluence/display/VCL/Apache+VCL.)

Google Drive lets the user store and access files anywhere—on the web, on the hard drive, or on the go. It works as follows:

- Go to Google Drive on the web at drive.google.com.
- Install Google Drive on the computer or mobile device.
- Store files in Google Drive. It is available on the device from which it is accessed.

After doing so, user will be able to access files from anywhere he or she wants to. If a file is changed on the web, by using a computer, or a mobile device, it is updated on every device where Google Drive is installed.

12.4.3 Google Docs

Another SaaS offering by Google is Google Docs. It is one of the many cloud computing document-sharing services. The majority of document-sharing services require user fees, whereas Google Docs is free. Its popularity among businesses is growing due to its enhanced sharing features and accessibility.

In addition, Google Docs has enjoyed a rapid rise in popularity among students and educational institutions.

Google Cloud Connect is a plug-in for Microsoft Office 2003, 2007, and 2010 on Windows that can automatically store and synchronize any Microsoft Word document, PowerPoint presentation, or Excel spreadsheet to Google Docs in Google Docs or Microsoft Office formats. The Google Doc copy is automatically updated each time the Microsoft Office document is saved. Microsoft Office documents can be edited offline and synchronized later when online. Google Cloud Sync maintains previous Microsoft Office document versions and allows multiple users to collaborate by working on the same document at the same time.

Google Spreadsheets and Google Sites also incorporate Google Apps Script to write code within documents in a similar way to Visual Basic for Applications (VBA) in Microsoft Office. The scripts can be activated either by user action or by a trigger in response to an event.

Google Forms and Google Drawings have been added to the Google Docs suite. Google Forms is a tool that allows users to collect information via a personalized survey or quiz. The information is then collected and automatically connected to a spreadsheet with the same name. The spreadsheet is populated with the survey and quiz responses.

Google Drawings allows users to collaborate creating, sharing, and editing images or drawings. Google Drawings contains a subset of the features in Google Presentation (Google Slides) but with different templates.

12.4.4 Dropbox

Dropbox is a file hosting service operated by Dropbox, Inc. that offers cloud storage, file synchronization, and client software. Dropbox allows users to create a special folder on each of their computers, which Dropbox then synchronizes so that it appears to be the same folder (with the same contents) regardless of which computer is used to view it. Files placed in this folder also are accessible through a website and mobile phone applications.

Dropbox provides client software for Microsoft Windows, Mac OS X, Linux, Android, iOS, BlackBerry OS, and web browsers, as well as unofficial ports to Symbian, Windows Phone, and MeeGo.

Both the Dropbox server and desktop client software are primarily written in Python. The desktop client uses GUI toolkits such as wxWidgets and Cocoa. Other notable Python libraries include Twisted, ctypes, and pywin32. Dropbox ships and depends on the librsync binary-delta library (which is written in C).

The Dropbox client enables users to drop any file into a designated folder that is then synchronized with Dropbox's Internet service and to any other of the user's computers and devices with the Dropbox client. Users may also upload files manually through a web browser.

Dropbox client supports synchronization and sharing along with personal storage. It supports revision history, so files deleted from the Dropbox folder may be recovered from any of the synced computers. Dropbox supports multiuser version control, enabling several users to edit and repost files without overwriting versions. The version history is by default kept for 30 days, with an unlimited version called *Pack-Rat* available for purchase.

Dropbox also provides a technology called LAN sync, which allows computers on a local area network to securely download files locally from each other instead of always hitting the central servers.

12.5 Open Source Tools for Research

The recent efforts to design and develop cloud technologies focus on defining novel methods, policies, and mechanisms for efficiently managing cloud infrastructures. To test these newly developed methods and policies, researchers need tools particularly simulators that allow them to evaluate the hypothesis prior to real deployment in an environment where one can reproduce tests. Simulation-based approaches in evaluating cloud computing systems and application behaviors offer significant benefits, as they allow cloud developers (i) to test performance of their provisioning and service delivery policies in a repeatable and controllable environment free of cost, and (ii) to tune the performance bottlenecks before real-world deployment on commercial clouds. Some of the tools that can be used for research work are explained in the following subsections.

12.5.1 CloudSim

CloudSim is a toolkit designed for creating a simulation environment for working in cloud. As a completely customizable tool, it allows extension and definition of policies in all the components of the software stack, which makes it suitable as a research tool that can handle the complexities arising from simulated environments.

Figure 12.6 shows the layered architecture of CloudSim. It explains the different layers involving UI, cloud services, and resources. Mapping of user code to the simulation environment is clearly shown in the layered architecture.

12.5.2 SimMapReduce

SimMapReduce is a simulator designed to be a flexible toolkit and is convenient to inherit or be inherited by other packages. SimMapReduce makes an effort to model a vivid MapReduce environment, considering some special features such as data locality and dependence between map and reduce, and it provides essential entity services that can be predefined in XML format. Furthermore, by using this simulator, modelers can realize multilayer scheduling algorithms on user level, job level, or task level by easily extending preserved classes.

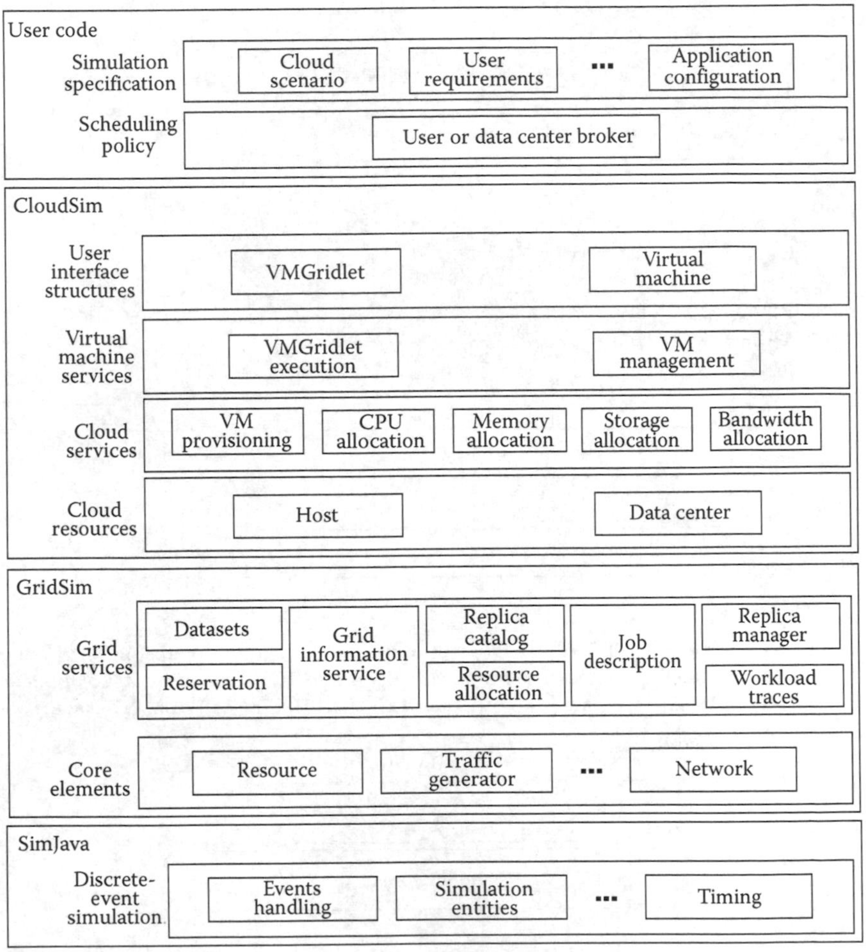

FIGURE 12.6
Layered CloudSim architecture. (Adapted from Buyya, R. et al., Modeling and simulation of scalable Cloud computing environments and the CloudSim toolkit: Challenges and opportunities, *International Conference on High Performance Computing and Simulation, 2009 (HPCS'09)*, IEEE, 2009, pp. 1–11.)

SimMapReduce is very useful for analyzing MapReduce programs in the simulation environment. Figure 12.7 gives the architectural diagram of SimMapReduce. It explains the mapping of user code with the simulation environment.

12.5.3 Cloud Analyst

Cloud Analyst is a CloudSim-based tool developed at the University of Melbourne whose goal is to support evaluation of social network tools according to geographic distribution of users and data centers. In this tool,

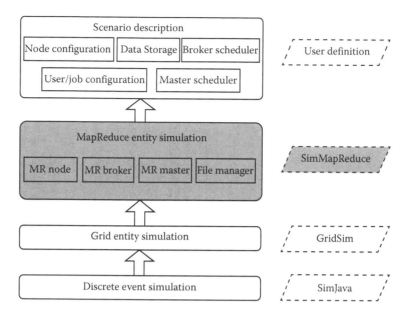

FIGURE 12.7
Four-layered architecture of SimMapReduce. (Adapted from Teng, F. et al., SimMapReduce: A simulator for modeling MapReduce framework, *2011 Fifth FTRA International Conference on Multimedia and Ubiquitous Engineering (MUE)*, June 28–30, 2011, IEEE, 2011, pp. 277–282.)

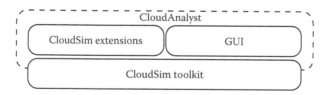

FIGURE 12.8
CloudAnalyst architecture. (Adapted from Wickremasinghe, B. et al., Cloudanalyst: A cloud-sim-based visual modeller for analysing cloud computing environments and applications, *2010 24th IEEE International Conference on Advanced Information Networking and Applications (AINA)*, IEEE, 2010, pp. 446–452.)

communities of users and data centers supporting the social networks are characterized based on their location; parameters such as user experience while using the social network application and load on the data center are obtained/logged. Figure 12.8 shows the architecture of the Cloud Analyst tool.

12.5.4 GreenCloud

GreenCloud is a sophisticated packet-level simulator for energy-aware cloud computing data centers with a focus on cloud communications. It offers a detailed fine-grained modeling of the energy consumed by the data center IT equipment, such as computing servers, network switches, and communication links. GreenCloud can be used to develop novel solutions in monitoring, resource allocation, workload scheduling, as well as optimization of communication protocols and network infrastructures. It can simulate existing data centers, guide capacity extension decisions, as well as help design future data center facilities.

GreenCloud, released under the General Public License Agreement, is an extension of the well-known NS2 network simulator. About 80% of the GreenCloud code is implemented in C++, while the remaining 20% is in the form of tool command language (TCL) scripts.

Figure 12.9 shows the architecture of the GreenCloud simulator. It presents the structure of the GreenCloud extension mapped onto the three-tier data center architecture.

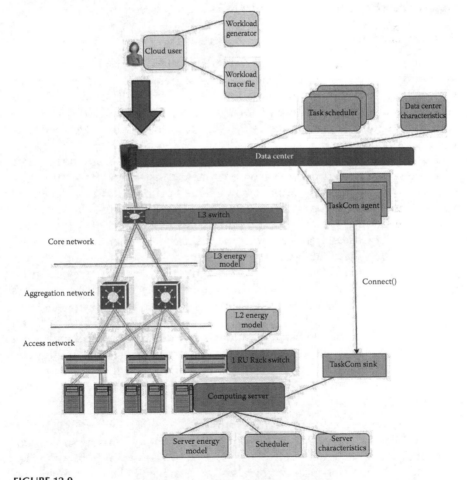

FIGURE 12.9
Architecture of the GreenCloud simulator. (Adapted from Kliazovich, D. et al., GreenCloud: A packet-level simulator of energy-aware Cloud computing data centers, *2010 IEEE Global Telecommunications Conference (GLOBECOM 2010)*, Miami, FL, December 6–10, 2010, pp. 1–5.)

12.6 Distributed Computing Tools for Management of Distributed Systems

A distributed computer system consists of multiple software components that are on multiple computers, but run as a single system. The computers that are in a distributed system can be physically close together and connected by a local network, or they can be geographically distant and connected by a wide area network. A distributed system can consist of any number of possible configurations, such as mainframes, personal computers, workstations, and minicomputers. The goal of distributed computing is to make such a network work as a single computer. The following subsections briefly explain some of the open source tools available for distributed computing.

12.6.1 Cassandra

Apache Cassandra is an open source–distributed database management system designed to handle large amounts of data across many commodity servers, providing high availability with no single point of failure. Cassandra offers robust support for clusters spanning multiple data centers, with asynchronous masterless replication allowing low-latency operations for all clients.

Cassandra also places a high value on performance. Cassandra's data model is a partitioned row store with tunable consistency. Rows are organized into tables; the first component of a table's primary key is the partition key; within a partition, rows are clustered by the remaining columns of the key. Other columns may be indexed separately from the primary key.

The Apache Cassandra database is the right choice when one needs scalability and high availability without compromising performance. Linear scalability and proven fault tolerance on commodity hardware or cloud infrastructure make it the perfect platform for mission-critical data. Cassandra's support for replicating across multiple data centers is the best in class, providing lower latency for the users, and tables may be created, dropped, and altered at runtime without blocking updates and queries. Cassandra does not support joins or subqueries, except for batch analysis via Hadoop. Rather, Cassandra emphasizes denormalization through features like collections.

12.6.2 Hadoop

The Apache Hadoop software library is a framework that allows for the distributed processing of large datasets across clusters of computers using simple programming models. It is designed to scale up from single servers to thousands of machines, each offering local computation and storage. Rather than relying on hardware to deliver high availability, the library itself is designed to detect and handle failures at the application layer, so delivering a highly available service on top of a cluster of computers, each of which may be prone to failures.

The Apache Hadoop framework is composed of the following modules:

- *Hadoop common*: It contains libraries and utilities needed by other Hadoop modules.
- *Hadoop distributed file system (HDFS)*: It is a distributed file system that stores data on the commodity machines, providing very high-aggregate bandwidth across the cluster.
- *Hadoop YARN*: It is a resource management platform responsible for managing compute resources in clusters and using them for scheduling of users' applications.
- *Hadoop MapReduce*: It is a programming model for large-scale data processing.

All the modules in Hadoop are designed with a fundamental assumption that hardware failures (of individual machines or racks of machines) are common and thus should be automatically handled in software by the framework. Apache Hadoop's MapReduce and HDFS components are originally derived, respectively, from Google's MapReduce and Google File System (GFS) papers.

Figure 12.10 shows the Hadoop architecture. It explains the working of a master node and slave nodes in the MapReduce framework. Master node has

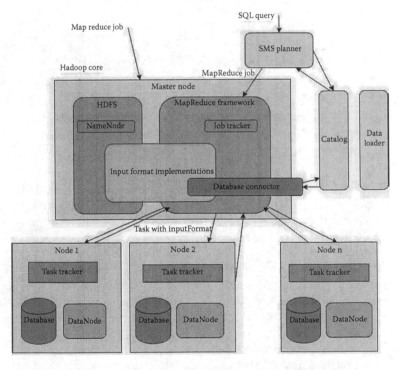

FIGURE 12.10
Hadoop architecture. (Adapted from http://hadoopdb.sourceforge.net/guide/.)

a job tracker and name node, whereas slave nodes have a task tracker and data node. Master node maps the work to slaves and later collects the result and reduces it to the required output. Slave nodes do the required computation and give the result to master.

12.6.3 MongoDB

MongoDB is another document database that provides high performance, high availability, and easy scalability. Documents (objects) map nicely to programming language data types. Embedded documents and arrays reduce the need for joins. Dynamic schema makes polymorphism easier. Embedding also makes reads and writes fast. Indexes can include keys from embedded documents and arrays achieving high performance. Replicated servers provide high availability in case of automatic master failover. Automatic sharding distributes collection data across machines ensures easy scalability. Eventually, consistent reads can be distributed over replicated servers.

MongoDB is a server process that runs on Linux, Windows, and OS X. It can be run both as a 32- and 64-bit application.

12.6.4 NGrid

NGrid is an open source grid computing framework written in C#. NGrid aims to be platform independent via the Mono project. It provides

- Transparent multithread programming model for grid programming
- Physical grid framework and some grid implementations
- Common utilities both for grid programming and grid implementations

NGrid, or precisely NGrid.Core, is a two-side grid abstraction layer for .Net. The first side is the grid programming model, and the second side is the grid services abstraction. NGrid.Core is not tightening to any particular grid. By implementing the grid services, NGrid.Core could be connected to any physical grid.

NGrid.Core provides a grid garbage collected (GC) model. In other words, objects are created and live in the grid. When not referenced anymore, objects are collected. Grid threads run on top of those grid objects. Synchronizations and communications between grid threads work in the same way as that of local threads.

The objective of any physical grid implementation is to manage as efficiently as possible this grid model. In order to make the optimization problem tractable, NGrid.Core includes some code metadata, known as attributes in C#, that could be used to decorate the client code. The NGrid. Core tuning attributes could be used by the physical grid in order to adapt the grid behavior for better performance. These attributes have an effect limited to the computational efficiency of the grid. In particular, they do not have any semantic impact: with or without tuning attribute, the result returned at the end of the grid computation is the same. The only parameter that may vary is the time of execution.

12.6.5 Ganglia

Ganglia is a scalable distributed monitoring system for high-performance computing systems such as clusters and grids. It is based on a hierarchical design targeted at federations of clusters. It leverages widely used technologies such as XML for data representation, XDR for compact, portable data transport, and RRD tool for data storage and visualization. It uses carefully engineered data structures and algorithms to achieve very low per-node overheads and high concurrency. The implementation is robust, has been ported to an extensive set of operating systems and processor architectures, and is currently in use on thousands of clusters around the world. It has been used to link clusters across university campuses and around the world and can scale to handle clusters with 2000 nodes.

12.7 Summary

In this chapter, we have discussed about various tools for cloud environment provided as an open source. We have categorized the tools into different groups like open source support for IaaS, PaaS, and SaaS, researchers, and distributed computing. Under each section, we have listed and briefly explained the architecture and features of some of the tools to provide useful information to the users. Table 12.1 gives the information related to service

TABLE 12.1

Open Source Service Providers

Provider Name	Service Model	Deployment Model	Server Operating System
GoGrid	IaaS	Public, private	Windows, Linux
OpenShift Origin	PaaS	Public, private	Linux
OpenNebula	IaaS	Public	Windows
Eucalyptus	IaaS	Public, private, hybrid	Windows, Linux
Cloudify	PaaS	Private	Linux
Paasmaker	PaaS	Private	Linux

model and deployment models offered by some providers. It also gives details of server operating systems. For further reading, papers and links given in reference section will be useful.

Review Points

- *Cluster controller* (CC): It is one of the main components of Eucalyptus architecture that manages node controllers and helps in deploying and managing instances on nodes (see Section 12.2.2).
- *Node controller* (NC): It is the basic component for all nodes that maintains the life cycle of instances running on each node (see Section 12.2.2).
- *Cloud controller* (CLC): It is the front end that provides web services interface to clients and also interacts with the rest of the components of Eucalyptus infrastructure (see Section 12.2.2).
- *Walrus storage controller* (WS3): It is a simple file storage system of Eucalyptus infrastructure that stores machine images and snapshots (see Section 12.2.2).
- *Storage controller* (SC): It is a storage system that provides persistent block storage. It allows creation of snapshots of volume (see Section 12.2.2).
- *Horizon*: It is the code name for *dashboard* in OpenStack architecture (see Section 12.2.3).
- *Nova*: It is the code name for *compute* in OpenStack architecture (see Section 12.2.3).
- *Keystone*: It is the code name for *identity* in OpenStack architecture (see Section 12.2.3).
- *Cinder*: It is the code name for *block storage* in OpenStack architecture (see Section 12.2.3).
- *Swift*: It is the code name for *object storage* in OpenStack architecture (see Section 12.2.3).
- *Neutron*: It is the code name for *network* in OpenStack architecture (see Section 12.2.3).
- *Glance*: It is the code name for *image* in OpenStack architecture (see Section 12.2.3).
- *Cartridges*: Cartridges represent pluggable components that can be combined within a single application (see Section 12.3.2).
- *Gear*: A gear represents the slice of the node's CPU, RAM, and base storage that is made available to each application (see Section 12.3.2).

- *Recipe*: A recipe is the execution plans or life cycle events (installing, starting, configuring, monitoring, upgrading and, etc.) of an application (see Section 12.3.4).
- *Cloud simulators*: These are tools that are useful for researchers to test the newly developed methods and policies and thereby evaluate the hypothesis (Section 12.5).

Review Questions

1. List the differences between open source tools and closed source tools.
2. Explain the main components of Eucalyptus.
3. What is OpenShift Origin? Explain.
4. Explain the architecture of OpenStack.
5. Explain the importance of open source tools for researchers.
6. What are the different open source tools for SaaS? Explain any of them.
7. What is MongoDB? Explain in detail.
8. Explain the architecture of the Hadoop framework.

References

1. http://blog.dsa-research.org/?author=7&paged=2.
2. http://mdshaonimran.wordpress.com/2011/11/26/eucalyptus-and-its-components/.
3. http://ken.pepple.info/openstack/2012/09/25/openstack-folsom-architecture/.
4. http://vcl.apache.org/info/architecture.html.
5. https://cwiki.apache.org/confluence/display/VCL/Apache+VCL.
6. Buyya, R., R. Ranjan, and R. N. Calheiros. Modeling and simulation of scalable Cloud computing environments and the CloudSim toolkit: Challenges and opportunities. *International Conference on High Performance Computing and Simulation, 2009 (HPCS'09)*. IEEE, 2009, pp. 1–11.
7. Teng, F., L. Yu, and F. Magoulès. SimMapReduce: A simulator for modeling MapReduce framework. *2011 Fifth FTRA International Conference on Multimedia and Ubiquitous Engineering (MUE)*, June 28–30, 2011. IEEE, 2011, pp. 277–282.

8. Wickremasinghe, B., R. N. Calheiros, and R. Buyya. Cloudanalyst: A cloudsim-based visual modeller for analysing cloud computing environments and applications. *2010 24th IEEE International Conference on Advanced Information Networking and Applications (AINA)*. IEEE, 2010, pp. 446–452.
9. Kliazovich, D., P. Bouvry, Y. Audzevich, and S. U. Khan. GreenCloud: A packet-level simulator of energy-aware Cloud computing data centers. *2010 IEEE Global Telecommunications Conference (GLOBECOM 2010)*, Miami, FL, December 6–10, 2010, pp. 1–5.
10. http://hadoopdb.sourceforge.net/guide/.

Further Reading

Beal, V. Cloud computing explained, 2012. http://www.webopedia.com/quick_ref/cloud_computing.asp. Accessed May 25, 2012.
The Apache Software Foundation. 2009. http://cassandra.apache.org.
The Apache Software Foundation. 2012. http://hadoop.apache.org. Accessed December 12, 2013.
Cordeiro, T., D. Damalio, N. Pereira, P. Endo, A. Palhares, G. Gonçalves, D. Sadok et al. Open source cloud computing platforms. *2010 9th International Conference on Grid and Cooperative Computing (GCC)*. IEEE, 2010, pp. 366–371.
de la Cruz, V. M. In a nutshell: How OpenStack works, 2013. Internet: http://vmartine-zdelacruz.com/in-a-nutshell-how-openstack-works. Accessed February 1, 2013.
Endo, P. T., G. E. Gonçalves, J. Kelner, and D. Sadok. A survey on open-source cloud computing solutions. *Brazilian Symposium on Computer Networks and Distributed Systems*, 2010.
Ganglia monitoring system. http://ganglia.sourceforge.net.
http://en.wikipedia.org/wiki/Apache_Cassandra. Accessed October 13, 2013.
http://siliconangle.com/blog/2013/08/21/10gen-boosts-hadoop-ecosystem-with-upgraded-connector/mongodb-architecture/. Accessed October 1, 2013.
http://www.mongodb.org/. Accessed October 3, 2013.
https://www.openshift.com/products. Accessed November 1, 2013.
Nurmi, D., R. Wolski, C. Grzegorczyk, G. Obertelli, S. Soman, L. Youseff, and D. Zagorodnov. The eucalyptus open-source cloud-computing system. *Ninth IEEE/ACM International Symposium on Cluster Computing and the Grid, 2009 (CCGRID'09)*. IEEE, 2009, pp. 124–131.
Paasmaker Team. Passmaker 0.9 documentation. http://docs.paasmaker.org/introduction.html. Accessed October 29, 2013.
Sempolinski, P. and D. Thain. A comparison and critique of Eucalyptus, OpenNebula and Nimbus. *2010 IEEE Second International Conference on Cloud Computing Technology and Science (CloudCom)*. IEEE, Indianapolis, IN, 2010, pp. 417–426.
Vermorel, J. NGrid Open source grid computing. http://ngrid.sourceforge.net/. Accessed September 10, 2013.

13

Security in Cloud Computing

Learning Objectives

The main objective of this chapter is to provide an overview of security issues in cloud computing. After reading this chapter, you will

- Understand the different security aspects
- Understand the security issues in cloud service models
- Understand identity management and access control issues related to security
- Understand audit and compliance in security

Preamble

Security is an important aspect to be considered in the cloud computing environment. This chapter focuses on different aspects of security. We begin with the introduction to cloud security. Subsequent sections talk about data security; virtualization security; security issues in Software as a Service (SaaS), Infrastructure as a Service (IaaS), and Platform as a Service (PaaS) models; etc. This chapter also considers privacy challenges and identity and access management issues in cloud. After reading this chapter, the reader can get an overview of security issues in different service models in cloud. The reader can also get an idea of challenges in these security issues.

13.1 Introduction

Cloud computing has entered everyone's life today irrespective of technology or any other aspect. Every tech magazine or every information technology (IT) organization website speaks about cloud computing. What exactly is this cloud computing? What does it do to make our lives more easier? To answer all these questions, let us look into the details of cloud computing with respect to technology more precisely.

13.1.1 Cloud in Information Technology

Cloud computing has revolutionized the IT industry for the past decade and is still developing creative ways to solve current problems. Companies and research institutes are slowly moving to the cloud to address their computing needs.

So, why is the cloud attractive to the public and private sectors? To answer this question, let us see what is put into cloud by an organization.

To make cloud more user friendly for computing, the industry has invested a lot into the following aspects:

1. *Time and finance*: The cloud is a centralized system and updates real-time information. Businesses with time-sensitive data are quick to grab this opportunity and harness the efficiency of the cloud. For example, medical researches that needed months of in-house number crunching moved to distributed systems, significantly reducing computing time and expenses.

2. *People and association*: With the advent of cloud, an online collaboration between distributed teams became easy. It is now easier to communicate and work with people located in different areas, sometimes different countries, during office hours. Teams now consist of members distributed across large geographic areas. As mentioned, the capability of the cloud to update information in real time enables teams to address issues immediately. Working together no longer means meeting up in the boardroom. Internet Protocol (IP) telephony, such as Skype and Google Hangout, provides a platform that allows team members to discuss tasks without stepping a foot outside their cubicles.

3. *Replacing hardware*: Relocating information and data systems to the cloud not only saves money but also reduces wasted resources. Companies no longer need to purchase hardware and systems that need installation and maintenance. Data centers on the cloud can reallocate these resources to clients by saving company dollars only by paying what is used and avoiding the purchase of machines that will not be useful in the long run. Cloud service providers (CSPs) also have the ability to optimize their systems to reduce waste. They also have the capability of upgrading their systems according to service demands. This is usually very expensive for businesses to do and results in wasted resources. Fewer in-house machines means that companies could redirect funds toward improving other aspects of business operations.

4. *Energy efficient*: A study reports that clients of Salesforce produced 95% less carbon compared to companies with systems in their premises.

5. *Study from Accenture, Microsoft, and WSP Environment and Energy*: A 2010 study from Accenture, Microsoft, and WSP Environment and Energy reported a huge impact of the cloud on CO_2 emissions. They found out that businesses with systems and applications on the cloud could reduce per-user carbon footprint by 30% for large companies and 90% for small businesses.

6. *Going green*: Greenpeace pointed out in a recent study that while efficiency is increasing, the energy source is also varying. The internal operations of data centers are *green*, but it is superficial if the power source is nonrenewable. With the increasing demand for cloud computing, energy consumption is expected to increase by 12% each year. An analysis of Greenpeace showed that out of the 10 leading tech companies—Akamai, Amazon, Apple, Facebook, Google, HP, IBM, Microsoft, Twitter, and Yahoo!—Akamai and Yahoo! are the most environment friendly and Apple the least. The report also highlighted Google's effort in *greening* its energy sources.

7. *The future of the cloud and the environment*: Two companies in Iceland, Green Earth Data and GreenQloud, both claim to offer 100% renewable energy by powering their data centers with geothermal and hydropower resources, which are abundant in the country. "The internet with cloud computing is becoming a big contributor to carbon emissions because of dirty energy usage," GreenQloud aims to set an example to cloud computing giants in creating environment-friendly cloud services. As the cloud industry expects to grow to a $150 billion market by the end of the year, users are increasingly demanding green services. Cloud technologies are quickly taking off, and it is a chance for companies and businesses to think of creative ways of harnessing its power while saving the environment.

13.1.2 Cloud General Challenges

The use of the cloud provides a number of opportunities like enabling services to be used without any understanding of their infrastructure. Cloud computing works using economies of scale. Vendors and service providers claim costs by establishing an ongoing revenue stream. Data and services are stored remotely but accessible from anywhere.

Though cloud is the hotcake technology today, there are many issues related with it. The following four major issues stand out with cloud computing:

1. *Threshold policy*: To test if the program works, develops, or improves and implements, a threshold policy is a pilot study before moving the program to the production environment. Check how the policy

detects sudden increases in the demand and results in the creation of additional instances to fill in the demand. Also, check to determine how unused resources are to be deallocated and turned over to other work.

2. *Interoperability issues*: The problems of achieving interoperability of applications between two cloud computing vendors. The need to reformat data or change the logic in applications.

3. *Hidden costs*: Cloud computing does not tell you what hidden costs are. In an instance of incurring network costs, companies who are far from the location of cloud providers could experience latency, particularly when there is heavy traffic.

4. *Unexpected behavior*: The tests to be made to show unexpected results of validation or releasing unused resources. The need to fix the problem before running the application in the cloud.

13.2 Security Aspects

Security concerns in the cloud are not that different from noncloud service offerings although they are exasperated—because in a single-tenant, noncloud environment, you generally know where information is and how it is being kept. There are many different customers and there is no mechanism followed to isolate each other's data.

Cloud computing places business data into the hands of an outside provider and makes regulatory compliance inherently riskier and more complex than it is when systems are maintained in-house. Loss of direct oversight means that the client company must verify that the service provider is working to ensure that data security and integrity are ironclad. The following are the current security-related research areas in cloud computing:

1. Reliable, distributed applications based on the Internet, such as the e-commerce system, rely heavily on the trust path among involved parties.

2. The skyrocketing demand for a new generation of cloud-based consumer and business applications is driving the need for a next generation of data centers that must be massively scalable, efficient, agile, reliable, and secure. In order to scale cloud services reliably to millions of service developers and billions of end users, the next-generation cloud computing and data center infrastructure will have to follow an evolution similar to the one that led to the creation of scalable telecommunication networks.

3. In the future, network-based CSPs will leverage virtualization technologies to be able to allocate just the right levels of virtualized compute, network, and storage resources to individual applications based on real-time business demand while also providing full service-level assurance of availability, performance, and security at a reasonable cost.

13.2.1 Data Security

Due to huge infrastructure, cost organizations are slowly switching to cloud technology. Data are stored in the CSP's infrastructure. As data do not reside in organization territory, many complex challenges arise. Some of the complex data security challenges in cloud include the following:

- The need to protect confidential business, government, or regulatory data
- Cloud service models with multiple tenants sharing the same infrastructure
- Data mobility and legal issues relative to such government rules as the European Union (EU) Data Privacy Directive
- Lack of standards about how CSPs securely recycle disk space and erase existing data
- Auditing, reporting, and compliance concerns
- Loss of visibility to key security and operational intelligence that no longer is available to feed enterprise IT security intelligence and risk management
- A new type of insider who does not even work for your company but may have control and visibility into your data

Such issues raise the level of anxiety about security risks in the cloud. Enterprises worry whether they can trust their employees or need to implement additional internal controls in the private cloud and whether third-party providers can provide adequate protection in multitenant environments that may also store competitor data.

There is also an ongoing concern about the safety of moving data between the enterprise and the cloud, as well as how to ensure that no residual data remnants remain upon moving to another CSP.

Unquestionably, virtualized environments and the private cloud involve new challenges in securing data, mixed trust levels, and the potential weakening of separation of duties and data governance. The public cloud compounds these challenges with data that are readily portable, accessible to anyone connecting with the cloud server, and replicated for availability. And with the hybrid cloud, the challenge is to protect data as it moves back and forth from the enterprise to a public cloud.

However, security and privacy are still cited by many organizations as the top inhibitors of cloud services adoption, which has led to the introduction of cloud encryption systems in the past 18 months.

The issues that must be addressed are as follows:

Breach notification and data residency: Not all data require equal protection, so businesses should categorize data intended for cloud storage and identify any compliance requirements in relation to data breach notification or if data may not be stored in other jurisdictions.

Gartner also recommends that enterprises should put in place an enterprise data security plan that sets out the business process for managing access requests from government law enforcement authorities. The plan should take stakeholders into account, such as legal, contract, and business units, security, and IT.

Data management at rest: Businesses should ask specific questions to determine the CSP's data storage life cycle and security policy.

Businesses should find out if

- Multitenant storage is being used, and if it is, find out what separation mechanism is being used between tenants
- Mechanisms such as tagging are used to prevent data being replicated to specific countries or regions

Storage used for archive and backup is encrypted and the key management strategy includes a strong identity and access management policy to restrict access within certain jurisdictions.

Gartner recommends that businesses use encryption to implement end-of-life strategies by deleting the keys to digitally shred the data while ensuring that keys are not compromised or replicated.

Data protection in motion: As a minimum requirement, Gartner recommends that businesses ensure that the CSP will support secure communication protocols such as Secure Socket Layer (SSL)/Transport Layer Security (TLS) for browser access or virtual private network (VPN)–based connections for system access for protected access to their services.

The research note says that businesses always encrypt sensitive data in motion to the cloud, but if data are unencrypted while in use or storage, it will be incumbent on the enterprise to mitigate against data breaches.

In IaaS, Gartner recommends that businesses favor CSPs that provide network separation among tenants, so that one tenant cannot see another's network traffic.

13.2.1.1 Data Center Security

Data are stored in outside territory of the user in a location called as *data center*, which is unknown to the user. As the location of the data center is unknown to the user, it becomes a virtual data center. The backbone of this virtual data center is virtual infrastructure, or the virtual machine (VM); however, virtual platforms are dependent on many other, often forgotten components of both the physical and virtual data centers.

There are typically seven areas of concern that accompany any major virtual platform implementation or migration. Often, these issues are not seen during staging and testing and only appear when the VMs take on the same amount of load as physical machines. The critical points represent two cornerstones of the data center: network and storage.

Lack of performance and availability: Virtualization moves many I/O (Input/Output) tasks tuned for hardware to software via the hypervisor. The virtualization translation layer is responsible for translating the optimized code for the software chip to the physical chip or CPU running on the underlying hardware. I/O intensive applications, like cryptographic processing applications for SSL, do not fare well when virtualized because of this translation layer.

VM saturation caused by virtualization sprawl can cause unanticipated resource constraints in all parts of the data center. With a physical machine running a network application, that application can have access to the full resources of the network card. This can lead to overall network performance issues, reduced bandwidth, and increased latency; the application might not be able to deal with all these issues. Even smaller issues such as IP address availability can be impacted by virtualization sprawl.

Lack of application awareness: One of the limitations of hypervisor- and kernel-based virtualization solutions is that they only virtualize the operating system (OS). OS virtualization does not virtualize nor is it even aware of applications that are running on the OS. Even the same applications do not realize that they are using virtual hardware on top of a hypervisor. By inserting this extra software management layer between the application and the hardware, the applications might encounter performance issues that are beyond control.

Virtual infrastructure platforms typically include software that can migrate live VM instances from one physical device to another; VMware Distributed Resource Scheduler (DRS) and VMotion are examples of live migration solutions. Like basic OS virtualization, these migration tools are unaware of the application state and also have no insight into the application delivery network.

Additional, unanticipated costs: Two of the primary drivers for virtualization are cost reduction and data center consolidation; however, implementing a complete VM solution in the data center and migrating physical machines to VMs do not come cheap. Once virtualization hardware and software are acquired, operational expenses can grow unbounded. Management of these new tools can be a long-term recurring cost, especially if the virtualization is done in-house. There can be additional growth requirements for the application and storage networks as these VMs begin to burden the existing infrastructure. Unexpected and unplanned costs can be a serious problem when implementing or migrating from physical to VMs, hindering or even completely halting deployment.

Unused virtualization features: New virtual platforms include many advanced networking technologies, such as software switching and support for virtual local area network (VLAN) segmentation. Often, new technologies perform flawless in development and staging, but they are unable to scale to production levels once deployed. These new platforms may have problems integrating with existing application and storage networks, requiring a redesign of the data center.

Storage integration issues tend to arise as soon as VMs are moved into production environments. First and foremost, network storage is a requirement for virtual platforms that implement live machine migration; direct-attached and local storage will only work for running local VMs. While many enterprise storage networks include technologies for data replication that span multiple geographic data centers, VM migration tools are often limited to local storage groups.

Overflowing storage network: Although converting physical machines to VMs is an asset for building dynamic data centers, hard drives become extremely large flat file virtual disk images. Consequently, file storage becomes unmanageable.

Congested storage network: Due to the portable nature of OS virtualization, there can be a dramatical increase in data traversing the storage network. For the same reason that VM disk files can overrun physical storage, once these images are made portable, it becomes trivial to move these VM images across the network from one host to another or from one storage array to another. It can be a challenge to prevent flooding of the storage network when planning a large VM migration or move. And as virtual sprawl and unmanaged VMs begin to appear in the data center, unplanned VM migrations can literally bring the network to a standstill, even on a LAN.

Management complexity: Throughout all areas of the data center, managing VMs as part of the complete management solution can be a struggle at best. VMs will report the metrics such as latency and response time of all physical machines. The management challenge with VMs appears in two forms:

1. The addition of two new components that need to be managed: The hypervisor and the host system. Neither one of these devices that exist in the physical server world is not part of existing data center management solutions, but they do have a major impact on the performance of VMs. Managing these devices and insight into these performance differences are critical.
2. Managing VMs, application network, and storage network together: Many VM platforms include built-in management tools, some of them highly sophisticated, such as VMware's Virtual Server.

While these tools do provide essential management tasks, such as live migration of virtual guests from one host to another, they do not take into account any external information. With physical servers, there is a line segregating ownership and management responsibilities. The network team owns the network fabric (switches, routers, VLANs, IPs), and the server team owns the servers (hardware, OS, application, uptime). OS virtualization blends these responsibilities onto one platform, blurring the lines of ownership and management.

13.2.1.2 Access Control

As the data are stored in the data center, accessing these critical data is a major concern. Being a web-based platform, the cloud acts according to the access rights reserved for the users to access the data. These access rights though well defined by individual firms still pose problems in the cloud. Gartner recommends that businesses require the CSP to support IP subnet access restriction policies so that enterprises can restrict end user access from known ranges of IP addresses and devices.

The enterprise should demand that the encryption provider offers adequate user access and administrative controls, stronger authentication alternatives such as two-factor authentication, management of access permissions, and separation of administrative duties such as security, network, and maintenance.

13.2.1.3 Encryption and Decryption

As the data are stored in cloud out of the territory of the user, it is recommended that users store data in encrypted form. Enterprises should always aim to manage the encryption keys, but if they are managed by a cloud encryption provider, enterprises must ensure that access management controls are in place that will satisfy breach notification requirements and data residency.

If keys are managed by the CSP, then businesses should require hardware-based key management systems within a tightly defined and managed set of key management processes. When keys are managed or available in the cloud, it is imperative that the vendor provides tight control and monitoring of potential snapshots of live workloads to prevent the risk of analyzing the memory contents to obtain the key.

Businesses should also require

Logging of all user and administrator access to cloud resources and to provide these logs to the enterprise in a format suitable for log management or security information and event management systems

The CSP to restrict access to sensitive system management tools that might *snapshot* a live workload, perform data migration, or back up and recover data

That images captured by migration or snapshotting tools are treated with the same security as other sensitive enterprise data.

13.2.2 Virtualization Security

Virtualization is technology that drives server consolidation and data center operations to a key ingredient in creating a flexible, on-demand infrastructure. When adopting virtualization for cloud computing, it becomes evident that the management tools used in a physical server-based deployment will not suffice in a highly dynamic virtualized one. To begin with, in a physical server deployment model, provisioning automation is generally not as heavily used unless there is a significant enough number of server OSs to warrant doing so.

Virtualization mainly focuses on three different areas: virtual networks (network virtualization), storage virtualization, and server virtualization. In network virtualization, the available resources are pooled into a network and the network bandwidth is split up into multiple channels where each individual channel is independent of one another. Storage virtualization combines the physical storage from multiple network storage devices, and this available storage is viewed as multiple different singular storage devices. In server virtualization, the identity of individual server devices is masked from the users, and the servers are designed to view as individual servers where the resource sharing and maintenance complexity are managed in a balanced way. The combination of these three virtualization components provides a self-managing capability to the resources, and this self-managing plays a major role in cloud computing.

The typical strategy for provisioning physical servers involves repetitive steps. In a heavily virtualized environment like the cloud, OS provisioning will rapidly transition to being a highly automated process. The critical areas of concern during virtualization are as follows.

A new threat: Virtualization alters the relationship between the OS and hardware. This challenges traditional security perspectives. It undermines the comfort you might feel when you provision an OS and application on a server you can see and touch. Some already believe that this sense of comfort is misplaced in most situations. For the average user, the actual security posture of a desktop PC with an Internet connection is hard to realistically discern.

Virtualization complicates the picture but does not necessarily make security better or worse. There are several important security concerns you need to address in considering the use of virtualization for cloud computing.

One potential new risk has to do with the potential to compromise a VM hypervisor. If the hypervisor is vulnerable to exploit, it will become a primary target. At the scale of the cloud, such a risk would have a broad impact if not otherwise mitigated. This requires an additional degree of network isolation and enhanced detection by security monitoring.

In examining this concern, first consider the nature of a hypervisor. It is observed that "Hypervisors are purpose-built with a small and specific set of functions. A hypervisor is smaller, more focused than a general

purpose operating system, and less exposed, having fewer or no externally accessible network ports. A hypervisor does not undergo frequent
change and does not run third-party applications. The guest operating
systems, which may be vulnerable, do not have direct access to the hypervisor. In fact, the hypervisor is completely transparent to network traffic
with the exception of traffic to/from a dedicated hypervisor management
interface."

Storage concerns: Another security concern with virtualization has to do with
the nature of allocating and deallocating resources such as local storage
associated with VMs. During the deployment and operation of a VM, data
are written into physical memory. If it is not cleared before those resources
are reallocated to the next VM, there is a potential for exposure.

These problems are certainly not unique to virtualization. They have been
addressed by every commonly used OS. Not all OSs manage data clear-
ing. Some might clear data upon resource release; others might do so upon
allocation.

The bottom line is to clear the data yourself, carefully handle operations
against sensitive data, and pay particular attention to access and privilege
controls. Another excellent security practice is to verify that a released
resource was cleared.

A further area of concern with virtualization has to do with the potential
for undetected network attacks between VMs collocated on a physical server.
Unless you can monitor the traffic from each VM, you cannot verify that traffic is not possible between those VMs.

In essence, network virtualization must deliver an appropriate network
interface to the VM. That interface might be a multiplexed channel with all the
switching and routing handled in the network interconnect hardware. Most
fully featured hypervisors have virtual switches and firewalls that sit between
the server physical interfaces and the virtual interfaces provided to the VMs.

Traffic management: Another theoretical technique that might have potential
for limiting traffic flow between VMs would be to use segregation to gather
and isolate different classes of VMs from each other. VMs could be traced
to their owners throughout their life cycle. They would only be colocated
on physical servers with other VMs that meet those same requirements for
colocation.

One actual practice for managing traffic flows between VMs is to use
VLANs to isolate traffic between one customer's VMs and another customer's VMs. To be completely effective, however, this technique requires
extending support for VLANs beyond the core switching infrastructure and
down to the physical servers that host VMs.

The next problem is scaling VLAN-like capabilities beyond their current
limits to support larger clouds. That support will also need to be standardized to allow multivendor solutions. It will also need to be tied in with network management and hypervisors.

13.2.3 Network Security

Cloud is based on networking of many things together like the network of infrastructure. While the network is the backbone of the cloud, many challenges are encountered in this network. Some of the challenges in the existing cloud networks are discussed in the following.

Application performance: Cloud tenants should be able to specify bandwidth requirements for applications hosted in the cloud, ensuring similar performance to on-premise deployments. Many tiered applications require some guaranteed bandwidth between server instances to satisfy user transactions within an acceptable time frame and meet predefined service-level agreements (SLAs). Insufficient bandwidth between these servers will impose significant latency on user interactions. Therefore, without explicit control, variations in cloud workloads and oversubscription can cause delay and drift of response time beyond acceptable limits, leading to SLA violations for the hosted applications.

Flexible deployment of appliances: Enterprises deploy a wide variety of security appliances in their data centers, such as deep packet inspection (DPI) or intrusion detection systems (IDSs), and firewalls to protect their applications from attacks. These are often employed alongside other appliances that perform load balancing, caching, and application acceleration. When deployed in the cloud, an enterprise application should continue to be able to flexibly exploit the functionality of these appliances.

Policy enforcement complexities: Traffic isolation and access control to end users are among the multiple forwarding policies that should be enforced. These policies directly impact the configuration of each router and switch. Changing requirements, different protocols (e.g., Open Shortest Path First [OSPF], LAG (Link Aggregation Group), Virtual Router Redundancy Protocol [VRRP]), and different flavors of L2 spanning tree protocols, along with vendor-specific protocols, make it extremely challenging to build, operate, and interconnect a cloud network at scale.

Topology-dependent complexity: The network topology of data centers is usually tuned to match a predefined traffic requirement. For instance, a network topology optimized for east–west traffic (i.e., traffic among servers in a data center) is not the same as a topology for north south (traffic to/from the Internet). The topology design also depends on how L2 and/or L3 is utilizing the effective network capacity. For instance, adding a simple link and switch in the presence of a spanning tree–based L2 forwarding protocol may not provide additional capacity. Furthermore, evolving the topology based on traffic pattern changes also requires complex configuration of L2 and L3 forwarding rules.

Application rewriting: Applications should run *out of the box* as much as possible, in particular for IP addresses and network-dependent failover mechanisms. Applications may need to be rewritten or reconfigured before deployment in the cloud to address several network-related limitations.

Two key issues are (1) lack of a broadcast domain abstraction in the cloud network and (2) cloud-assigned IP addresses for virtual servers.

Location dependency: Network appliances and servers (e.g., hypervisors) are typically tied to a statically configured physical network, which implicitly creates a location-dependent constraint. For instance, the IP address of a server is typically determined based on the VLAN or the subnet to which it belongs. VLANs and subnets are based on physical switch port configuration. Therefore, a VM cannot be easily and smoothly migrated across the network. Constrained VM migration decreases the level of resource utilization and flexibility. Besides, physical mapping of VLAN or subnet space to the physical ports of a switch often leads to a fragmented IP address pool.

Multilayer network complexity: A typical three-layer data center network includes a TOR (Top of Rack) layer connecting the servers in a rack, an aggregation layer, and a core layer, which provides connectivity to/from the Internet edge. This multilayer architecture imposes significant complexities in defining boundaries of L2 domains, L3 forwarding networks and policies, and layer-specific multivendor networking equipment. Providers of cloud computing services are currently operating their own data centers. Connectivity between the data centers to provide the vision of *one cloud* is completely within the control of the CSP. There may be situations where an organization or enterprise needs to be able to work with multiple cloud providers due to locality of access, migration from one cloud service to another, merger of companies working with different cloud providers, cloud providers who provide the best-of-class services, and similar cases. Cloud interoperability and the ability to share various types of information between clouds become important in such scenarios. Although CSPs might see less immediate need for any interoperability, enterprise customers will see a need to push them in this direction. This broad area of cloud interoperability is sometimes known as cloud federation.

13.3 Platform-Related Security

CSPs offer services in various service models like SaaS, PaaS, and IaaS where the user is offered varied services based on his or her requirements. Every service model offered brings with it many security-related challenges like secured network, locality of resources, accessing secure data, data privacy, and backup policy.

13.3.1 Security Issues in Cloud Service Models

Cloud computing uses three delivery models such as SaaS, PaaS, and IaaS through which different types of computing services are provided to the end user. These three delivery models provide infrastructure resources, application platform, and software as services to the cloud customer. These service models place a different level of security requirement in the cloud environment. IaaS is the basis of all cloud services, with PaaS built upon it and SaaS in turn built upon it. Just as capabilities are inherited, so are the information security issues and risks. There are significant trade-offs to each model in terms of integrated features, complexity versus extensibility, and security. If the CSP takes care of only the security at the lower part of the security architecture, the consumers become more responsible for implementing and managing the security capabilities.

SaaS is a software deployment model in which applications are remotely hosted by the application or service provider and made available to customers on demand, over the Internet. The SaaS model offers the customers with significant benefits, such as improved operational efficiency and reduced costs. SaaS is rapidly emerging as the dominant delivery model for meeting the needs of enterprise IT services. SaaS is rapidly emerging as the dominant delivery model for meeting the needs of enterprise IT services. PaaS is one layer above IaaS on the stack and abstracts away everything up to OS, middleware, etc. This offers an integrated set of developer environment that a developer can tap to build their applications without having any clue about what is going on underneath the service. It offers developers a service that provides a complete software development life cycle management, from planning to design to building applications to deployment to testing to maintenance. Everything else is abstracted away from the *view* of the developers.

13.3.2 Software-as-a-Service Security Issues

In a traditional on-premise application deployment model, the sensitive data of each enterprise continue to reside within the enterprise boundary and are subject to its physical, logical, and personnel security and access control policies. The architecture of SaaS-based applications is specifically designed to support many users concurrently (multitenancy). SaaS applications are accessed through the web, and so web browser security is very much important. Information security officers will need to consider various methods of securing SaaS applications. Web services (WS) security, Extensible Markup Language (XML) encryption, SSL, and available options used in enforcing data protection transmitted over the Internet. In the SaaS model, the enterprise data are stored outside the enterprise boundary, at the SaaS vendor end. Consequently, the SaaS vendor must adopt additional security checks to ensure data security and to prevent breaches due to security vulnerabilities in the application or through malicious employees. This involves the use of strong encryption techniques for data security and fine-grained authorization to control access to data. The pain points of concern in SaaS are as follows.

Network security: In an SaaS deployment model, sensitive data flow over the network needs to be secured in order to prevent leakage of sensitive information. This involves the use of strong network traffic encryption techniques such as the SSL and TLS for security.

Resource locality: In an SaaS model of a cloud environment, the end users use the services provided by the cloud providers without knowing exactly where the resources for such services are located. Due to compliance and data privacy laws in various countries, locality of data is of utmost importance in much enterprise architecture.

The directive prohibits transfers of personal data to countries that do not ensure an adequate level of protection. For example, the recent Dropbox users have to agree to the *Terms of Service* that grant the providers the right to disclose user information in compliance with laws and law enforcement requests.

Cloud standards: To achieve interoperability among clouds and to increase their stability and security, cloud standards are needed across organizations. For example, the current storage services by a cloud provider may be incompatible with those of other providers. In order to keep their customers, cloud providers may introduce so-called sticky services that create difficulty for the users if they want to migrate from one provider to the other.

Data segregation: Multitenancy is one of the major characteristics of cloud computing. In a multitenancy situation, data of various users will reside at the same location. Intrusion of data of one user by another becomes possible in this environment. This intrusion can be done either by hacking through the loop holes in the application or by injecting client code into the SaaS system. An SaaS model should, therefore, ensure a clear boundary for each user's data. The boundary must be ensured not only at the physical level but also at the application level. The service should be intelligent enough to segregate the data from different users.

Data access: Data access issue is mainly related to security policies provided to the users while accessing the data. The organizations will have their own security policies based on which each employee can have access to a particular set of data. The security policies may entitle some considerations, wherein some of the employees are not given access to a certain amount of data. These security policies must be adhered by the cloud to avoid intrusion of data by unauthorized users. The SaaS model must be flexible enough to incorporate the specific policies put forward by the organization. The model must also be able to provide organizational boundary within the cloud because multiple organizations will be deploying their business processes within a single cloud environment.

Data breaches: Since data from various users and business organizations lie together in a cloud environment, breaching into the cloud environment will potentially attack the data of all the users. Thus, the cloud becomes a high-value target.

Backup: The SaaS vendor needs to ensure that all sensitive enterprise data are regularly backed up to facilitate quick recovery in case of disasters. Also, the use of strong encryption schemes to protect the backup data is recommended to prevent accidental leakage of sensitive information. In the case of cloud vendors such as Amazon, the data at rest in S3 are not encrypted by default. The users need to separately encrypt their data and backups so that it cannot be accessed or tampered with by unauthorized parties.

Identity management (IdM) and sign-on process: IdM deals with identifying individuals in a system and controlling the access to the resources in that system by placing restrictions on the established identities. When an SaaS provider has to know how to control who has access to what systems within the enterprise, it becomes all the more challenging task. In such scenarios, the provisioning and deprovisioning of the users in the cloud become very crucial.

13.3.3 Platform-as-a-Service Security Issues

PaaS provides a ready-to-use platform, including OS that runs on vendor-provided infrastructure. As the infrastructure is of the CSP, various security challenges of the focused architecture are caused mainly by the spread of the user objects over the hosts of the cloud. Stringently allowing access of objects to the resources and defending the objects against malicious or corrupt providers reasonably reduce possible risks. Network access and service measurement bring together concerns about secure communications and access control. Well-known practices, object scale enforcement of authorization, and undeniable traceability methods may alleviate the concerns.

Apart from the aforementioned problems, user privacy must be protected in a public, shared cloud. Therefore, proposed solutions must be privacy aware. Service continuity is another concern for many enterprises that consider cloud adoption. Accordingly, fault-tolerant reliable systems are required.

13.3.4 Infrastructure-as-a-Service Security Issues

Cloud computing makes a lot of promises in the areas of increased flexibility and agility, potential cost savings, and competitive advantages for developers so that they can stand up an infrastructure quickly and efficiently to enable them to develop the software to drive business success. There are a lot of problems that cloud, especially private cloud, solves, but it is not that much good in solving problems related to security.

However, in a private cloud environment, some of the traditional problems faced are as follows:

1. *Hypervisor security*: In private cloud, most or all of services will run in a virtualized environment and the security model used by the hypervisor cannot be taken for granted. A need to evaluate the security models and the development of hypervisors becomes necessary.

2. *Multitenancy*: Although all the tenants in the multitenancy environment will be from the same company, not all tenants may be comfortable sharing infrastructure with other users within the same company.

3. *Identity management and access control (IdAM)*: In a traditional data center, we were comfortable with the small handful of authentication repositories we had to work with—Active Directory being one of the most popular. But with private cloud, handling authentication and authorization for the cloud infrastructure, handling tenants, and handling delegation of administration of various aspects of the cloud fabric are the major tasks to be addressed.

4. *Network security*: In private cloud, we are likely to have many components of a service communicate with each other over virtual network channels only. Assessing the traffic, employing some powerful access controls for physical networks, and control quality of service, which is a key issue in the *availability* aspect of the confidentiality, integrity and availability (CIA) security model, are major concerns.

As a consolidation, platform-related security considers the previously said three service delivery models like SaaS, PaaS, and IaaS, and the concerned components are also mentioned individually.

Combining the three types of clouds (public cloud, private cloud, and hybrid cloud) together with the three service delivery models, we get a complete picture of a cloud computing environment interlinked by connectivity devices coupled with information security components. Virtualized physical resources, virtualized infrastructure, virtualized middleware platforms, and business-related applications are being provided as computing services in the cloud. Cloud providers and cloud consumers must be able to maintain and establish computing security at all levels of interfaces in the cloud computing architecture.

13.4 Audit and Compliance

It is a widely known fact that data protection and regulatory compliance are among the top security concerns for chief information officers (CIOs) of any organization.

According to the Pew Internet and American Life Project, an overwhelming majority of users of cloud computing services expressed serious concern about the possibility of a service provider disclosing their data to others. Ninety percent of cloud application users said that they would be very concerned if the company at which their data were stored sold them to another party.

A survey conducted by many firms expressed the view that security is the biggest challenge for the cloud computing model. Stakeholders, therefore, increasingly feel the need to prevent data breaches. In recent months, many newspaper articles have revealed data leaks in sensitive areas such as the financial and governmental domains and web community.

One of the missions of the data protection authorities is to prevent the so-called *Big Brother* phenomenon, which refers to a scenario whereby a public authority processes personal data without adequate privacy protection. In such a situation, end users may view the cloud as a vehicle for drifting into a totalitarian surveillance society.

The specificities of cloud computing, therefore, make the data protection incentive even greater. For example, the cloud provider should provide encryption to protect the stored personal data against unauthorized access, copy, leakage, or processing.

In a cloud environment, companies have no control over their data, which, being entrusted to third-party application service providers in the cloud, could now reside anywhere in the world. Nor will a company know in which country its data reside at any given point in time. This is a central issue of cloud computing that conflicts with the EU requirements whereby a company must at all times know where the personal data in its possession are being transferred to. Cloud computing thus poses special problems for multinationals with specific EU customers.

13.4.1 Disaster Recovery

Simple data backup as well as more comprehensive disaster recovery and business continuity planning is an essential part of business and personal life. Backup as a Service and Disaster Recovery as a Service is now available online through the cloud for every level of user, from personal, small business to large enterprise data storage and retrieval, either publicly through the Internet or via more secure dedicated access methods. As a result, traditional methods are becoming obsolete.

A few of the advantages include the following:

- No huge upfront costs for capital investment or infrastructure management or black boxes.
- Backups are physically stored in a different location from the original source of your data.
- Remote backup does not require user intervention or periodic manual backups.
- Unlimited data retention. You can get as much or as little data storage space as you need.
- Backups are automatic and *smart*. They occur continuously and efficiently back up your files only as the data change.

Cloud computing, based on virtualization, takes a very different approach to disaster recovery. With virtualization, the entire server, including the OS, applications, patches, and data, is encapsulated into a single software bundle or virtual server. This entire virtual server can be copied or backed up to an offsite data center and spun up on a virtual host in a matter of minutes.

Since the virtual server is hardware independent, the OS, applications, patches, and data can be safely and accurately transferred from one data center to a second data center without the burden of reloading each component of the server. This can dramatically reduce recovery times compared to conventional (nonvirtualized) disaster recovery approaches where servers need to be loaded with the OS and application software and patched to the last configuration used in production before the data can be restored.

13.4.2 Privacy and Integrity

The promise to deliver IT as a service is addressed to a large range of consumers, from small- and medium-sized enterprises (SMEs) and public administrations to end users. Users are creating an ever-growing quantity of personal data.

This expanding quantity of personal data will drive demand for cloud services, particularly if cloud computing delivers on the promises of lower costs for customers and the emergence of new business models for providers.

Among the main privacy challenges for cloud computing are as follows.

Complexity of risk assessment in a cloud environment: The complexity of cloud services introduces a number of unknown parameters. Service providers and consumers are cautious about offering guarantees for compliance-ready services and adopting the services. With service providers promoting a simple way to flow personal data irrespective of national boundaries, a real challenge arises in terms of checking the data processing life cycle and its compliance with legal frameworks.

To address the issues like stakeholders' roles and responsibilities, data replication, and legal issues compliance, the Madrid Resolution states that every responsible person shall have transparent policies with regard to the processing of personal data. Stakeholders need to specify requirements for cloud computing that meet the expected level of security and privacy. In Europe, the European Network and Information Security Agency (ENISA) provides recommendations to facilitate the understanding of the shift in the balance of responsibility and accountability for key functions such as governance and control over data and IT operations and compliance with laws and regulations.

Emergence of new business models and their implications for consumer privacy: A report by the Federal Trade Commission (FTC) on *Protecting consumer privacy in an era of rapid change* analyzes the implications for consumer privacy of technological advances in the IT sphere. According to the FTC, users are able to collect, store, manipulate, and share vast amounts of consumer data for very little cost.

These technological advances have led to an explosion of new business models that depend on capturing consumer data at a specific and individual level and over time, including profiling, online behavioral advertising (OBA), social media services, and location-based mobile services.

13.5 Summary

Cloud being the efficient, low-cost computing platform for the IT industry is fast growing. With the tremendous growth comes the challenge of handling the critical data and offering quality of service to the users.

This chapter throws light on various security aspects related to cloud computing. The basic security elements related to various cloud deployment models and service delivery models are briefly explained here. The security aspects like data center security and security with respect to service models are also highlighted. As the IT industry is driven to the cloud for its computing capacities, it in turn needs to look into the security of the critical data stored in third-party providers. The various security-related issues, though addressed by IT industry, still are major concerns as no standard development procedure is defined for the development of the cloud model. As organizations follow their own model for development, security becomes more prominent for concentration.

Review Points

- *Threshold policy*: To test if the program works, develops, or improves and implements, a threshold policy is a pilot study before moving the program to the production environment (see Section 13.1.2).

- *Data security*: Data are stored in the CSP's infrastructure. As data do not reside in organization territory, many complex challenges arise (see Section 13.2.1).

- *Access control*: As the data are stored in the data center, accessing these critical data is a major concern. Being a web-based platform, the cloud acts according to the access rights reserved for the users to access the data (see Section 13.2.1.2).

- *Location dependency*: Network appliances and servers (e.g., hypervisors) are typically tied to a statically configured physical network, which implicitly creates a location-dependent constraint (see Section 13.2.3).

- *Resource locality*: In an SaaS model of a cloud environment, the end users use the services provided by the cloud providers without knowing exactly where the resources for such services are located (see Section 13.3.2).

- *IdM*: IdM deals with identifying individuals in a system and controlling the access to the resources in that system by placing restrictions on the established identities (see Section 13.3.2).

- *Disaster recovery*: Simple data backup as well as more comprehensive disaster recovery and business continuity planning is an essential part of business and personal life (see Section 13.4.1).

Review Questions

1. What are the issues to be addressed in data security? Explain.
2. What are the storage concerns in virtualization security?
3. Explain the challenges in cloud networks.
4. What are the security issues in SaaS? Explain.
5. What are the security issues in PaaS? Explain.
6. What are the security issues in IaaS? Explain.
7. What are the advantages of Disaster Recovery as a Service?
8. What are the privacy challenges for cloud computing? Explain.

Further Reading

An introduction to cloud computing in public sector. White Paper, APPTIS.

Bioh, M. and D. Earhart. Security issues that affect cloud computing data storage. www.slideshare.net, 2009. Accessed November 2, 2014.

Brodkin, J. Gartner: Seven cloud-computing security risks, 2008. www.infoworld.com/d/security-central/gartner-seven-cloud-computing-security-risks-853. Accessed November 23, 2014.

Cloud computing issues and impacts. White Paper, Ernst and Young.

Cloud computing security and privacy issues. White Paper, CEPIS.

Curran, K., S. Carlin, and M. Adams Security issues in cloud computing. *Journal of Network Engineering* 4069–4072, 2011.

Hamlen, K., M. Kantarcioglu, L. Khan, and B. Thuraisingham. Security issues for cloud computing. *International Journal of Information Security and Privacy* 4(2): 39–51, 2010.

Hashizume, K., D. G. Rosado, E. Fernández-Medina, and E. B. Fernandez. An analysis of security issues for cloud computing. *Journal of Internet Services and Applications* 4: 5, 2013.

Cloud Computing, http://www.gartner.com/technology/research/cloud-computing/report. Accessed December 21, 2013.

http://www.hostway.com/resources/media/disaster-recovery-in-the-cloud.pdf. Accessed November 27, 2013.

Kuyoro, S. O., F. Ibikunle, and O. Awodele. Cloud computing security issues and challenges. *International Journal of Computer Networks (IJCN)* 3(5): 247–255, 2011.

Ma, M. and C. Meinel. A proposal for trust model: Independent trust intermediary service (ITIS). *Proceedings of IADIS International Conference WWW/Internet 2002,* Lisbon, Portugal, 2002, pp. 785–790.

More, J. J. Cloud computing: Information technology's answer to sustainability. www.ecoseed.org. Accessed November 9, 2013.

Murphy, A. Keeping your head above the cloud: Seven data center challenges to consider before going virtual. White Paper.

Rashmi, G. Sahoo, and S. Mehfuz. Securing software as a service model of cloud computing: Issues and solutions. *International Journal on Cloud Computing: Services and Architecture (IJCCSA)* 3(4): 1–11, 2013.

Subashini, S. and V. Kavitha. A survey on security issues in service delivery models of cloud computing. *Journal of Network and Computer Applications* 34(1): 1–11, 2011.

The ethics and security of cloud computing. White Paper, ILTA.

www.trustedcomputinggroup.org.

Xu, K., M. Song, X. Zhang, and J. Song. A cloud computing platform based on P2P. *Proceedings of the IEEE,* 2009.

14

Advanced Concepts in Cloud Computing

Learning Objectives

The objectives of this chapter are to

- Give an insight on some of the advanced topics in cloud
- Give a brief description about each topic
- Discuss the current problems on a particular topic
- Indicate some of the works that are going on in that area
- Discuss the benefits of using the technologies

Preamble

Cloud computing is an important area of research and is developing at a rapid pace. There are certain topics in cloud where there are some research openings. There are several applications of cloud; depth- and breadthwise, the cloud is expanding. In this chapter, several advanced topics are discussed. A brief description about intercloud and its types, topologies, and benefits is provided first, followed by cloud management. Cloud management is considered to be one of the important areas, and here certain management topics related to architecture are discussed. Consequently, mobile cloud and media cloud are discussed in brief. This is followed by a discussion on standards and interoperability. Here, the problems related to interoperability are pointed out. The necessity for standardization is also discussed. This is followed by a brief introduction to cloud governance, its various components, and its importance. The next subsection discussed in detail about green computing. It elaborates on the method for energy calculation. It discusses some of the popular algorithms that try to reduce the energy consumption of data center. Further, the recent works in this area are mentioned. The next subsection discusses about the impact of computational intelligence in cloud computing. The areas where it is used and the places where it can be used are discussed. Finally, cloud data analytics, one of the most popular topics that is changing today's world, is discussed. This subsection elaborates upon data analytics, and the reason cloud is used here. The chapter ends with some review points, review questions, and further readings.

14.1 Intercloud

Cloud computing is a technology that has been around for quite some time and has been a welcomed technology. The National Institute of Standards and Technology (NIST) defines cloud computing as "a model for enabling ubiquitous, convenient, on-demand network access to a shared pool of configurable computing resources (e.g., networks, servers, storage, applications, and services) that can be rapidly provisioned and released with minimal management effort or service provider interaction." Nowadays, there is an increased demand for cloud computing. As the number of cloud users increases, it is a challenge to cater to the requirements of all the users in order to maintain the credibility of the cloud providers. Cloud providers do not offer an infinite amount of resources and hence may get saturated at some point in time. In some cases, situations may arise where the cloud provider might not be able to satisfy the requirements of the customers. In such cases where the cloud provider is faced with an unexpected increase in requirements or need, it has to resort to some methods to ensure customer satisfaction. It is in such situations that intercloud comes as a boon to the cloud provider.

Intercloud can be basically viewed as a *cloud of clouds*. Here, multiple cloud providers join hands to serve the customers. Intercloud can take any of the two forms: federation of clouds or multicloud. The concept of intercloud was introduced way back in 2010 in the first *IEEE International Workshop on Cloud Computing Interoperability and Services (InterCloud 2010)*. In October 2013, an environment that would serve as the testbed for intercloud-related applications was proposed.

The major cloud providers are Amazon, Google, Microsoft, IBM, etc., as shown in Figure 14.1. It is a challenge for other cloud providers to compete with these cloud providers. Federation of cloud is an advantage to the minor cloud providers as they can provide their resources for rent and thus attain profit for their resources, which would otherwise be left underutilized.

A basic classification of intercloud can be given as in Figure 14.2.

In federation of clouds, the cloud providers manage the interconnections among them. Here, infrastructures may be shared so as to enable resource sharing. The users need not bother about using more than one cloud as the

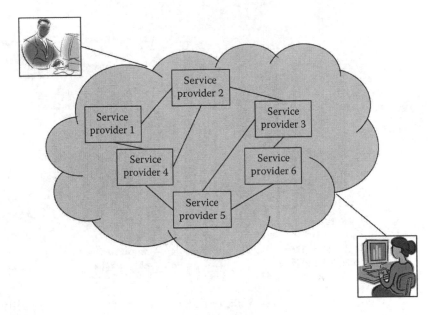

FIGURE 14.1
An illustration of intercloud. (Adapted from http://computersight.com/software/cloud-computing-chapter-one-review-and-guide/.)

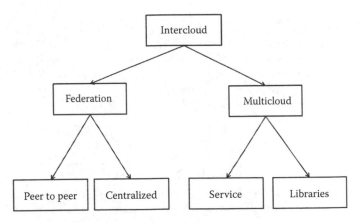

FIGURE 14.2
Classification of intercloud. (Adapted from http://2.bp.blogspot.com/.)

cloud providers take up the responsibility of providing a transparent service to the customers. Here, one cloud provider can rent the resources of another cloud provider and offer it to the customers.

In multicloud, the client or service makes use of multiple clouds. Thus, the user is aware of the fact that they are being served by more than one

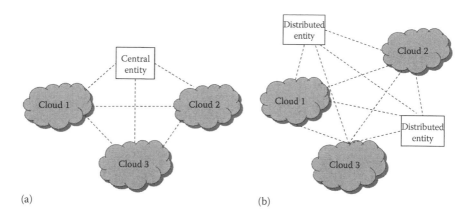

FIGURE 14.3
Topologies of the different intercloud architectures. (a) Centralized intercloud federation,
(b) peer to peer intercloud federation. (Adapted from http://cloudstrategypartners.blogspot.
in/2013/04/intercloud-not-all-same-federation.html.)

cloud provider. It is the responsibility of the users to provide interoperability
between the various cloud providers. The different types of multicloud and
federation cloud have various topologies, as shown in Figure 14.3.

Intercloud provides customers with a wide range of benefits such as the
following:

1. *Diverse geographical locations*: Cloud service providers may have their
   data centers in diverse geographic locations. In cases where the
   customers are bound by legislative constraints regarding the location of where their data are stored, the data center provided by the
   cloud service provider in a particular location may not be enough.
   Intercloud can be used to enable the customers to control the location of storage of their data.

2. *Better application resilience*: The objectives of fault tolerance, reliability, and availability can also be better fulfilled by the cloud providers
   making use of intercloud. By placing data in more than one data
   center, the applications can be made more fault tolerant. In case one
   of the data centers does not function, there is always an option to
   use an alternative copy at another data center. This also increases
   the availability. In case of abrupt increases in traffic, performance
   degradation may occur. This can be avoided in intercloud. Thus, performance is guaranteed.

3. *Avoidance of vendor lock-in*: The use of intercloud avoids vendor lock-
   in where the customer becomes dependent on a particular vendor.
   Here, users depend on multiple vendors.

4. *Flexibility*: By making use of intercloud, the users are assigned
   resources in a more flexible manner.

5. *Power saving*: In intracloud scenarios, energy can be conserved by making use of virtual machine (VM) migrations. Virtual machines can be transferred from overloaded hosts to underloaded hosts. The energy saving can be done at a higher level in intercloud scenarios where migrations can be carried out across data centers.

Intercloud brings along with it a new set of challenges that have to be resolved. The Internet is a network of networks. Coming up with standards for the Internet so that the various networks can cooperate with each other took a period of about 5 years. In a similar manner, various issues have to be tackled before interclouds can be used. The various cloud providers should have some mechanisms to establish trust among themselves. Other issues include SLA requirements and billing rates. Some cloud providers may have authentication mechanisms in order to offer their resources. In simple terms, the cloud providers should *match* with each other in terms of resources, policies, technologies, etc. Various existing technologies, such as Extensible Messaging and Presence Protocol (XMPP), may be used to realize the intercloud among heterogeneous cloud providers. Other issues include scalability, support of VM migrations across cloud providers, resource migration, security, policy management, and monitoring.

InterCloud is a cloud access operator that acts as a gateway to a variety of Software-as-a-Service (SaaS) platforms among intercloud partners, such as Microsoft Azure and Amazon Web Services. It provides access to any outsourced application platform in a secure manner.

There are various architectures proposed in the literature that aim to solve the various issues related to intercloud such as data exchange and resource sharing.

14.2 Cloud Management

As cloud computing gains wide acceptance among users, there is a need for management of cloud so as to ensure the proper functioning of the cloud services. The cloud management is responsible for managing the infrastructure of the cloud. It is the solution to the problem of realizing the expected economic benefits, by employing proper capabilities of management.

The term cloud management is the name given to the collection of software and technologies that is used to govern and monitor the various cloud applications. Cloud management ensures that the cloud services are running optimally and that it interacts with its coapplications. The cloud management software may need to handle heterogeneous resources. It has to monitor the various tasks such as resource allocation.

The strategies for cloud management include regular monitoring and auditing services, and initiating and managing plans for disaster recovery.

Cloud management usually provides a portal for customers. It provides user authentication, encryption, and budget management on behalf of various enterprises.

There are various cloud management interfaces available such as the enStratus Enterprise Cloud Management Solution, which provides the provisioning, management, and automation of application in major private and public clouds. One of the cloud management reference architecture introduced by the Distributed Management Task Force (DMTF) is shown in Figure 14.4.

The functional components of the architecture are as follows:

1. *Cloud service developer*: Designs, implements, and maintains service templates.
2. *Cloud service consumer*: Provides access to services for service users.
3. *Cloud service provider*: Supplies cloud services to internal or external consumers.
4. *Data artifacts*: Control and status elements exchanged across the provider interface.
5. *Provider interface*: Interface that allows consumers to access and monitor the contracted services.
6. *Profiles*: Specification that defines the associations, methods, and properties for a management domain.

The cloud management interface should take special care of security. If an attacker gains access to the cloud management interface, the aftereffects can be drastic.

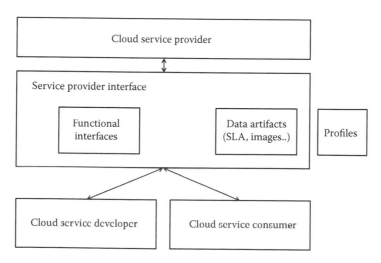

FIGURE 14.4
Cloud service reference architecture.

14.3 Mobile Cloud

The Mobile Cloud Computing Forum defines mobile cloud computing (MCC) as follows [4]:

> Mobile Cloud Computing at its simplest, refers to an infrastructure where both the data storage and the data processing happen outside of the mobile device. Mobile cloud applications move the computing power and data storage away from mobile phones and into the cloud, bringing applications and mobile computing to not just smartphone users but a much broader range of mobile subscribers.

MCC is basically the intersection of the fields of cloud computing and mobile networks. Mobile networks are basically networks that connect mobile users. The emergence of ultrafast mobile networks makes it necessary to bring the cloud domain to the mobile networking domain. This field is still in its primary stage of development. MCC basically enables the building and hosting of mobile applications over the cloud. There are various issues to be resolved such as live VM migration, security, privacy preserving, and fault tolerance. There is a probability that migrating VMs become an overhead in MCC.

Mobile devices usually have limited computing power and resources. The limitations of mobile computing can be overcome by MCC. It will allow users to access platforms and applications provided by the cloud through their mobile devices. Here, the users will no longer be restricted by the limited resources they own. Thus, more computing-intensive mobile applications can be supported by a larger number of devices.

There are many mobile cloud apps, the most common of which are Google's Gmail and Google Voice for iPhone. A general diagram is shown in Figure 14.5.

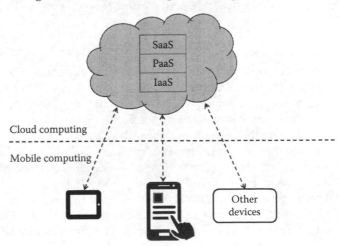

FIGURE 14.5
Mobile cloud computing.

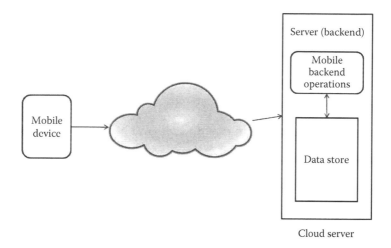

FIGURE 14.6
A remote cloud server offering services to a mobile device.

In mobile applications that involve image processing, natural language processing, multimedia search, and so on, the lack of resources on the mobile device can be handled by renting services offered by the cloud. A remote cloud server offering services is shown in Figure 14.6.

Another scenario is where mobile devices can themselves become part of a cloud and offer their resources for rent to other mobile devices. Thus, collective resources can be made available, provided that they fall within the vicinity. Another scenario makes use of cloudlets. Cloudlets are basically a collection of multicore computers connected to remote cloud servers. Mobile devices usually act as thin client devices to the cloudlet.

One major issue in MCC is the process of handing over the jobs to the cloud. This greatly depends on the distance that physically separates the cloud and the mobile device. This process might incur additional costs. These costs also have to be taken into consideration. There are many works in the literature dedicated to the cost–benefit analysis of MCC.

Another major issue that does not exist in traditional cloud computing is the support of user mobility. There should be mechanisms to identify the current location of the mobile client. The mobility of the users should not affect the connectivity to the cloud.

Energy efficiency is another aspect that requires intense research in the context of MCC. Certain architectures have been proposed for MCC. They must take into consideration the privacy, security, and trust issues, among other issues.

The benefits of MCC are as follows:

1. *Extending battery lifetime*: As far as smartphone users are concerned, battery life is one of the major problems. Though smartphones have a wide range of functionalities, the duration of battery lifetime drastically decreases when more workload is submitted to the smartphone. By offloading the execution of applications that require intensive computation to the cloud, the battery lifetime can be improved.

2. *Improving data storage capacity and processing power*: Having dealt with battery lifetime, the next common issue encountered by smartphone users is the storage capacity. Cloud provides a simple solution for this problem where the data can be accessed via wireless networks.

3. *Improving reliability*: While using cloud, the data get backed up on a number of other devices. In the case of accidental loss of data, the backed up data can be accessed.

4. *Improving scalability*: By using cloud, the scalability of mobile applications can also be easily improved.

MCC finds applications in mobile commerce, mobile learning, mobile healthcare, and many more.

14.4 Media Cloud

Multimedia data are the major form of data available in the Internet nowadays. In the case of videos, there are a wide variety of formats available. Storing and processing multimedia data require a large amount of processing and storage facilities. The currently emerging technology, cloud computing, can be utilized to efficiently process and store such data. Media cloud provides storage for media data and also enables the presentation of data using media signaling protocols. A simple multimedia cloud is shown in Figure 14.7.

Media clouds provide distributed processing of multimedia data and provide services with high Quality of Service (QoS) for multimedia data. The content stored in media cloud can be easily streamed to various clients such as music players in cars and smartphones, using protocols such as transmission control protocol (TCP), user diagram protocol (UDP), and real-time transport protocol (RTP).

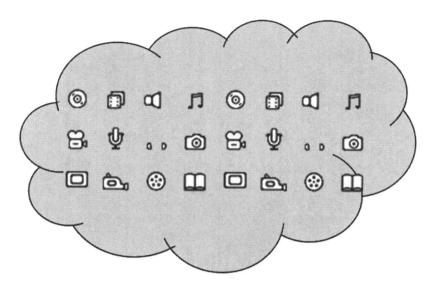

FIGURE 14.7
Multimedia cloud.

The process of streaming includes the activities of buffering, rendering, recording, and mixing of data. The users of media cloud will also be provided with an easier way to share multimedia data among them.

There are many challenges to the media cloud. The challenges come in the form of heterogeneous formats available for the different types of multimedia data, heterogeneity of applications, scalability to adapt to newly developed formats of media, and at the same time making the multimedia cloud a profitable cloud. For increasing the profit, the risk of failure should be decreased.

A reference architecture of the media cloud is presented in Figure 14.8.

The architecture in Figure 14.8 shows the basic services that should be provided by a media cloud: storage and infrastructure management, cluster and grid management, workflow automation, etc. This architecture contains five main components:

1. Cloud administrative services
2. Ingest services that accept media input from a wide range of sources
3. Streaming services
4. Video services that manage and deliver videos across media channels to various clients
5. Storage subsystems for content cache and movement, storage, and asset management

The only solution to the bursty increase in multimedia in the coming years will be cloud computing. The cloud will be the major host for all multimedia

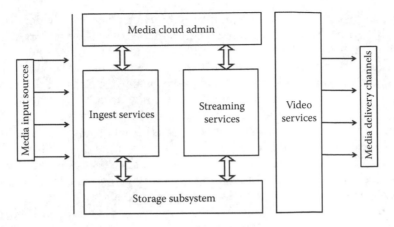

FIGURE 14.8
Functional components of media cloud architecture. (Adapted from Poehlein, S. et al., Moving to the media cloud, 2010, available [online]: http://h20195.www2.hp.com/V2/GetPDF. aspx/4AA2–1000ENW.pdf.)

data and multimedia processing in the near future. In order to make the systems less complex, an approach that transforms the heterogeneous types of audio, video, and other multimedia data into a standard format is more desirable. The area of media cloud has various unresolved issues such as media cloud architecture, storage, delivery of multimedia data, mobile broadcast, energy efficiency, and media mining that involves information retrieval from multimedia data. The media cloud can also be used for provisioning to meet Video on Demand requirements.

14.5 Interoperability and Standards

Interoperability is defined as the ability to move the workload from one cloud provider to another without any problem. Thus, a user should be able to use the cloud providers interchangeably or should be able to shift their work base from one cloud provider to another. Interoperability is one of the prime issues in cloud computing. One of the problems associated with interoperability is vendor lock-in problem.

Whenever a cloud service provider does not allow the users to migrate to other cloud, it is called as vendor lock-in. In simple terms, the vendor locks the user from moving from one vendor to another. The vendor lock-in has been an issue for a long time. Here, once a user has chosen one cloud service provider, then he will not be able to change or migrate to another service provider without starting his work again.

Another issue with interoperability is that the service providers do not allow the users to use products or components of other vendors in their cloud infrastructure. Thus, a user has to use the components given by one cloud provider only.

There are two possible reasons for this problem. One is that the vendors do not want to lose their customers, and so they do not allow them to use other vendor's components or to migrate to other vendors. Another is that there is very less technical support or there is very less advancement in this area, so even if the cloud components are open source or are not governed by any such policy or rule, there is very less chance that two clouds would be interoperable. Interoperability and portability go hand in hand. There are three aspects of portability, namely application portability, platform portability, and infrastructure portability. Thus, if the portability issues are resolved, then it would not take much time to make the clouds interoperable as there are only a few more issues that need to be considered other than portability.

All the aforementioned problems are due to one primary reason, and that is standardization. The main problem with the cloud service providers is that they do not have a common standard for all the cloud service providers. Each service provider follows their own standards, but these standards vary from company to company. Thus, no two companies have the same standards. According to a report by Carnegie Mellon University (CMU), there are different aspects of standardization on each service model of cloud. There are three basic service models of cloud: Infrastructure as a Service (IaaS), Platform as a Service (PaaS), and Software as a Service (SaaS). The standardization will affect all the three models in a different manner. According to CMU, the SaaS models would be the least benefitted by the standards. This is because such software involves licensing terms, and standardization would not have a greater effect on it. Similarly, PaaS also may not be that much benefitted from the standardization models. But, IaaS would have a considerable amount of benefits as the workloads or resources are in the form of VMs and storage disks, and if standardization is made, then the users can migrate their VMs from one service provider to another. There is a lot of effort being put by different organizations to make standards. There have been several attempts by several communities to have interoperable systems and to maintain the standards [6]. This process will continue in the future, and the day is not far when interoperability would be the property of every cloud service provider.

14.6 Cloud Governance

Governance is a term in the corporate world that generally involves the process of creating value to organization by creating strategic objectives that will lead to growth of the company and will maintain a certain level of control over the company. Governance is not to be confused with management. Both terms, though similar, have lot of difference. Governance comes into picture where there is a booming industry that involves a lot of resources including people. Governance involves maintaining and following certain policies throughout the company. These involve high-level decision making that would affect all the people related to the company.

Cloud computing is one of the rapidly growing technologies. Almost all the big companies have started using cloud. As the days go by, cloud is becoming bigger and vast. Not only computing resources but also the number of people working in cloud is increasing. Hence, there is a necessity of governance mechanism to maintain the growth of cloud. There is a need to regulate all the resources such that the people involved or affected by the cloud directly or indirectly are benefitted and all the things that happen through the cloud are being monitored properly. Cloud governance is very important because it is a known fact that well-governed organizations have a higher probability of sustaining business and retaining their position in the industry. If governed properly, an organization is able to adapt to the changes quickly.

Cloud governance involves certain hurdles. There are several responsibilities involved in cloud governance, which include some of the points by He [7]:

1. *Quality of service*: Quality of service is another major issue as far as an organization is concerned. There are no standard metrics or a standard way to ensure the quality of service to the customers. There are several models or algorithms that are proposed to ensure the quality of service to the users. Service-level agreements (SLA) are the major parameter that is considered for assuring QoS. These SLAs are considered to be very important. Negotiations are made between the cloud service provider and the cloud user based on these SLAs. An SLA gives flexibility to the cloud user, and a cloud user can claim or demand justice or compensation if SLA terms are violated in any way by the service provider. Thus, it is the responsibility of the service provider to ensure that the SLA terms are satisfied at any cost.

2. *Complying with the laws and standards*: This is one of the most difficult parts as far as cloud is concerned. As the back-end servers, storage disks and VMs are spread all over the world there are lots of issues being raised because of the data stored in the cloud. The problem comes when people from one country try to access the data stored in another country. The laws of both the countries may vary. Hence, coming to a conclusion on which law should be applied to which user and eventually what should be the standard law for the company is very difficult.

3. *Adapt to changing service mechanisms*: There are different service mechanisms involved in delivering services. Time to time, these service mechanisms change and the service provider should have the capability to adapt to different service mechanisms as quickly as possible.

4. *Data privacy issue*: The laws related to data privacy had been an issue for a long time. The data of the user stored should not be viewed by any outsider or even the service provider. There is no surety on whether the data details stored by the user are safe. Thus, this is one of the most important issues that need to be resolved.

5. *Multitenancy*: Multitenancy is a property of cloud that involves sharing of workspace (instance) by many users (clients). It is one of the important properties of cloud that made it popular. Though it is good, it has some issues to security. A user can break the security barrier and can access others' data, as all the users share a single system. Giving security to the user is still an issue that needs to be addressed.

6. *Governance inside the organization (internal)*: The rules inside the organization for people related to cloud projects should be given. As these projects are large and complex, the rules should be specified clearly. The number of people working on cloud and the number of resources being used for cloud in a company are very high. Hence, there needs to be a law related to these issues, that is, the way to use these resources should also be defined.

Based on the these issues, researchers have proposed some governance models like the Lifecycle Governance model and the Lewam Woldu Cloud Governance model. As cloud governance is similar to governance in service-oriented architecture (SOA) and thus both of them are considered as the subset of IT governance. This area is one of the emerging areas.

There are several aspects of governance that can be automated, and several companies have come up with a tool for cloud governance. These tools include cloud governance tool by Dell tool name Agility platform by service mesh etc. An active research is going on in this area, and companies and researchers are coming up with several models and tools for governance.

14.7 Computational Intelligence in Cloud

According to Andries P. Engelbrecht, computational intelligence (CI) is the study of adaptive mechanisms to enable or facilitate intelligent behavior in complex and changing environments. These mechanisms include those artificial intelligence (AI) paradigms that exhibit an ability to learn or adapt to new situations, generalize, abstract, discover, and associate [8]. An important aspect of CI is adaptivity. This field exists for nearly four decades. Scientists have been using these algorithms for solving different kinds of problems. CI is basically used for problems that cannot be solved by using conventional algorithms. There are many real-time problems that cannot be solved using conventional methods, and in these cases, some practical natural approaches prove to be a better choice. CI includes all the heuristic algorithms that are inspired by natural processes. Some of the popular examples are genetic algorithm and swarm optimization. Genetic algorithms come under the subdivision called evolutionary algorithms. These evolutionary algorithms are inspired by evolutionary process of the living being. Similarly, swarm optimization also comes under this category. Swarm optimization techniques are inspired by fish school or bird flocks, that is, their group behavior property is used. Artificial immune system is one of the methods based on our immune system. Similarly, there are several other methods that are based on natural processes. These computational methodologies are extensively used for optimization-related problems. This approach has an extensive use in cloud computing.

Cloud computing in simple terms is offering service (resources) to the customers. It involves a lot of resources at the back end. Managing resources is a difficult task. This resource management involves several important tasks like resource scheduling, provisioning, consolidation, and migration. Some of the aforementioned tasks like scheduling cannot have exhaustive methods. The prime ways of scheduling resources are workflow scheduling and task scheduling. Scheduling problem as such is a nondeterministic polynomial-time hard (NP-hard) problem. NP-hard problems are difficult problems that do not have an exhaustive solution and can only have approximate solutions. So for these kinds of problems, optimization techniques such as CI algorithm can be used. Optimization techniques are usually characterized by properties that give good approximate solutions that are near to the correct solution. The aforementioned are some of the instances. These algorithms consume less amount of time than the exhaustive search algorithms. The more the problems become complex, the more are the chances of using CI algorithms.

Advantages

- Suitable for complex problems for which exhaustive search is not possible
- Consumes less time than the exhaustive counterpart

14.8 Green Cloud

Data centers are the backbone of cloud computing. The data centers housing the clouds often use up a lot of energy. Data centers are the facility that houses computer systems and their related components. The power consumed by data centers mainly constitutes of the power required to run the actual equipment and the power used up by devices to cool the equipment. To cool the systems in the data center, we usually make use of precision air conditioners that control the temperature and humidity throughout the day and can also be managed remotely.

As more and more people switch to cloud computing, the energy consumed by it becomes significant. In an era of great concern for a *greener environment*, this area in cloud computing should also be considered. It is from this basic need that green cloud computing arose. Green cloud computing is basically the computing solutions to the problem of energy consumption. These solutions also aim at reducing the OPerational EXpenses (OPEX). The increase in operational costs reduces the marginal profit of cloud service providers. The power consumption of data centers also has a nonnegligible impact on the increase in carbon emissions and global warming. Thus, there is a need to develop energy-efficient solutions to cloud computing. For this, an in-depth analysis of power efficiency of clouds needs to be done (Figure 14.9).

The architecture of data centers is of different types such as two tier and three tier. In all these architectures, energy efficiency has to be analyzed. The data center efficiency is usually calculated by considering the performance delivered per watt. We have two metrics:

Power usage effectiveness (*PUE*): It compares the energy used for computing against the energy used for cooling and other overhead. The ideal value is 1.0. It can be calculated as

$$PUE = \frac{\text{Total facility energy}}{\text{IT equipment energy}}$$

The lowest PUE that has been achieved in the present scenario is 1.13 in Google data center.

Data center infrastructure efficiency (*DCiE*): This is the reciprocal of PUE:

$$DCiE = \frac{1}{PUE}$$

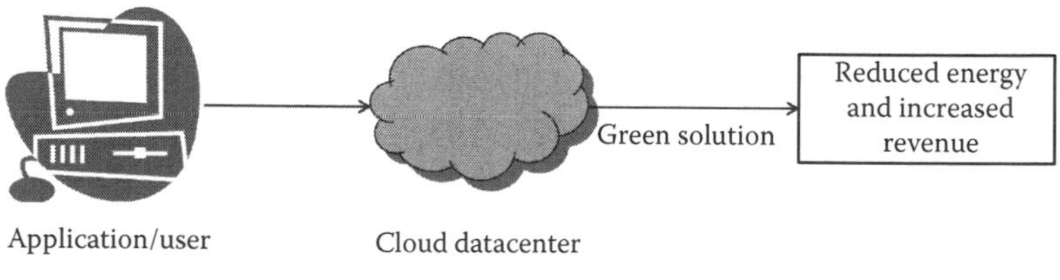

FIGURE 14.9
Green cloud computing. (Adapted from Garg, S.K. and Buyya, R., Green cloud computing and environmental sustainability, in *Harnessing Green IT: Principles and Practices,* S. Murugesan and G.R. Gangadharan (eds.), 2012, pp. 315–340.)

Studies have shown that idle servers consume around 66% of energy compared to their fully loaded configuration. The management of memory modules and other factors contribute to this energy consumption. Thus, we can save a considerable amount of power if the workloads are concentrated on a minimum number of the computing servers, thus enabling the shutting down of idle servers. Basically, the different power management techniques in computer architecture are as follows:

1. *Dynamic voltage scaling (DVS)*: Increasing or reducing the voltage in a component.
2. *Dynamic frequency scaling (DFS)*: This is also called CPU throttling, where the frequency of the microprocessor is adjusted so as to save power or to lessen the amount of generated heat.
3. *Dynamic shutdown (DNS)*: This scheme selectively shuts down the idle or underutilized components.

The green lining in cloud computing is provided by virtualization. Virtualization is creating a virtual entity. Using virtualization, multiple users can be accommodated on a single host. If there was no virtualization, each user would have to be allocated separate physical machines. This can be avoided by creating isolated virtual entities for each user on the same physical machine. Thus, virtualization is a tool for efficient utilization of resources. It can also serve as a tool for energy saving. Workload consolidation and server consolidation can be used to reduce energy consumption. When servers are overloaded, the VMs on the server can be transported to other underloaded servers. If there are many underloaded servers, the VMs can be consolidated to one or more servers, enabling the idle servers to be switched off.

There are various simulators to simulate the cloud computing environment. But not many of them measure the energy efficiency of cloud. To tackle this issue, the GreenCloud Simulator was developed in 2013 by a team led by Dzmitry Kliazovich at the University of Luxembourg. It provides a fine-grained simulation of energy-efficient clouds [10]. The simulator focuses on the communication between the various elements in the cloud. This approach is adapted as more than 30% of the total energy is consumed by the elements for communication.

There are various works in the literature that fall under the area of green cloud computing. There are energy-efficient scheduling strategies that make use of VM migrations. Green routing aims to provide routing service in an energy-efficient manner. Green networking is the area that focuses on energy-efficiency issues in networking. Another notable work in this area is the GreenCloud Architecture proposed by Garg and Buyya [9]. They propose a framework that curbs the energy consumption of clouds. They propose architecture with an intermediate layer called the GreenBroker. This GreenBroker makes use of a directory that contains the carbon emission details of various cloud providers so as to provide the *greenest* provider to the customer at the same time maximizing the profit. The carbon emission information of various cloud providers at IaaS level is measured by using energy meters. At the PaaS level, energy profiling tools are used.

The energy-saving strategies for data centers can be divided into four broad sections:

1. Energy-efficient techniques for servers
2. Energy-efficient techniques for network
3. Energy-efficient techniques for servers and network
4. Cooling and renewable energy

Under the fourth category, a notable work has been done by Baikie and Hosman [11] where an attempt is made to reduce the energy used by data centers. This includes methods such as using renewable energy to power the data centers and so on.

14.9 Cloud Analytics

Cloud analytics is a process of doing any business or data-intensive analysis (data analytics) in public or private cloud. Data analytics is a process of examining unprocessed or raw data to make some meaningful conclusion from the data. The results may include either a value or a set of values or a graph. Data analytics has been very popular. Several researchers all over the world are using this technique effectively, and based on the results, important decisions are taken. If it is done on large amount of data that can be in the range of terabytes and petabytes, it is called as big data analytics. Big data analytics has been a recent buzzword in the industry. Nowadays, almost the entire world is using the Internet, and the amount of data that are generated per day is very large.

There are several places and several companies that analyze some parts of these data to get a meaningful conclusion. For example, a social networking site can be considered. Usually, the numbers of users of a social networking site are large. Suppose the site administrator wants to analyze the users according to the game in which they are interested, he or she can do that by analyzing their whole data based on one or more parameters that determine or estimate the results. There are several other places apart from the Internet where a large amount of data are generated, and these data require a complete analysis. For example, if an insurance company wants to find out the best customer and wants to rate their customers based on their monthly payments, they can do this by using analytics. Similarly, there are several applications of data analytics.

Analyzing a small amount of data is easy, but when the amount of data is increased, it becomes difficult. The data analytics operation that is done in today's world is on data that are in terabytes and petabytes. For doing this kind of data-intensive operations, heavy back-end resources, and most importantly high-processing power, are needed. This would cost companies or organizations a lot. The companies or organizations sometimes cannot afford to buy that many resources, as buying and maintaining it are a big issue. In this case, there is no other option left for the organization other than renting the resources. There had been situations before when this kind of power was needed. For example, the Large Hadron Collider (LHC), which was used by CERN, recorded the data in petabytes per week and there was a necessity to process these data. The scientists did not have their own resources, so they relied upon the grid technology for the resources. The grid computing model is one of the computing technologies that allow the user to use the large number of resources on a pay-per-use basis. This technology is primarily intended for research organizations to use the vast processing resources available. These resources were provided for a fee, and primarily these were used for high-end scientific applications in areas of astronomy, physics, bioinformatics, etc. These are sometimes called utility grids and are considered as the reason for the success of many high data-intensive projects.

Similar to grid, cloud can also be used for renting resources. Cloud computing is a computing model where primarily resource, softwares, and platforms are offered as services. There is a major difference between grid and cloud computing. Unlike grid, cloud services are divided into several types.

In this context, two major types are private and public. Public cloud offers services to all the people around the globe, whereas private cloud is restricted and can only be used by an organization or individual. When the private or public cloud is used for data analysis, then it is called cloud analytics. Cloud analytics is very popular as it is analogous to grid platform and has an advantage that anybody in the world can use this platform for doing analytics.

People can use these public or private cloud services for analyzing the huge amount of data, and they can pay according to what they use; there is no need to buy any resource. This is very useful for companies in real time, because these can be used whenever it is necessary, that is, only when it is required and there is no necessity to maintain any hardware resources. Figure 14.10 depicts the classification of cloud analytics.

Another aspect of cloud analytics is that instead of using the private cloud or public cloud as such, people can use the cloud applications that are built in the cloud and that are designed specifically for data analysis. Thus, instead of using resources directly, the delivery model called SaaS is used. Here, cloud-based software is created by the provider and the user can analyze the data by using the software, which is usually a web application. This kind of cloud application will allow the user to use any kind of device that can access a simple web application.

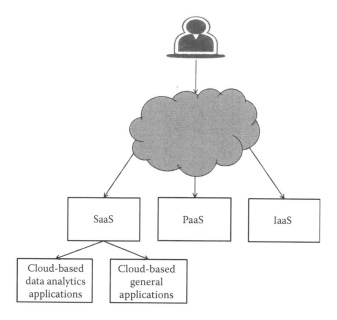

FIGURE 14.10
Cloud analytics.

14.10 Summary

This chapter briefly described the advanced topics in cloud computing. The topics that were included are related to recent advancements in this field. All the topics that are discussed have a potential impact. The topics either are important as an application or have certain research importance. The topics that were discussed briefly were mobile cloud, intercloud, media cloud, interoperability and standards, cloud governance, green cloud, cloud analytics, and impacts of CI in cloud computing and cloud management. All the aforementioned topics are discussed with illustrations on the benefits of each technology and their impact on the industry or academia.

Review Points

- *Federation*: Federation of clouds is the deployment and management of multiple cloud computing services to match business needs (see Section 14.1).

- *Multicloud*: Multicloud involves two or more cloud service providers (see Section 14.1).
- *Peer-to-peer intercloud federation*: This is one of the types of cloud federation where there is no central entity (see Section 14.1).
- *Centralized intercloud federation*: This is a type of cloud federation that involves a central entity (see Section 14.1).
- *Precision air conditioners*: These are the type of air conditioners used for cooling in data centers (see Section 14.8).
- *OPerational EXpenses (OPEX)*: OPerational EXpenses is the ongoing cost for running a product, business, or system (see Section 14.8).
- *Dynamic voltage scaling (DVS)*: This is the process of dynamically adjusting the voltage levels used in a data center (see Section 14.8).
- *Dynamic frequency scaling (DFS)*: This involves dynamically varying the frequency of the microprocessor (see Section 14.8).
- *Dynamic shutdown (DNS)*: This is a power management technique where idle servers are shut down (see Section 14.8).
- *Server consolidation*: This is the process of reducing the number of underutilized servers (see Section 14.8).
- *Cloudlet*: A cloudlet can be basically viewed as a mini data center (see Section 14.3).
- *Vendor lock-in*: In vendor lock-in problem, the cloud service provider does not allow the users to migrate to other cloud (see Section 14.5).
- *Cloud analytics*: Cloud analytics is a process of doing any business or data-intensive analysis (data analytics) in public or private cloud (see Section 14.9).
- *Multitenancy*: Multitenancy is a property of cloud that involves sharing of workspace (instance) by many users (clients) (see Section 14.6).
- *Data analytics*: Data analytics is a process of examining unprocessed or raw data to make some meaningful conclusion from the data (see Section 14.9).

Review Questions

1. What are the different types of intercloud topologies?
2. What are the differences between multicloud and federation of clouds?
3. Enumerate the various advantages of intercloud.

4. What are the different techniques for power management in computer architecture?

5. Give an outline of the energy-saving strategies for data centers.

6. Define the metrics used to measure the data center efficiency.

7. Discuss the different scenarios where MCC can be employed.

8. What are the major issues in multimedia cloud?

9. What is the need for cloud management?

10. Why standardization is necessary?

11. What are the advantages of using CI in cloud?

12. Explain vendor lock-in problem.

13. Why cloud governance is necessary?

14. What are the two ways of using cloud analytics?

15. What is an SLA?

References

1. http://computersight.com/software/cloud-computing-chapter-one-review-and-guide/. Accessed March 23, 2013.
2. http://2.bp.blogspot.com/. Accessed March 23, 2013.
3. http://cloudstrategypartners.blogspot.in/2013/04/intercloud-not-all-same-federation.html.
4. http://www.cse.wustl.edu/~jain/cse574–10/ftp/cloud/index.html. Accessed February 24, 2014.
5. Poehlein, S., V. Saxena, G. T. Willis, J. Fedders, and M. Guttmann Moving to the media cloud. Available [Online]: http://h20195.www2.hp.com/V2/GetPDF.aspx/4AA2–1000ENW.pdf (accessed August 15, 2010).
6. The role of standards in CloudComputing interoperability. Available [Online]: https://resources.sei.cmu.edu/asset_files/TechnicalNote/2012_004_001_28143.pdf. Accessed February 24, 2014.
7. He, Y. The lifecycle process model for cloud governance, 2011.
8. Engelbrecht, A. P. *Computational Intelligence: An Introduction*, 2nd edn. Wiley Publications.
9. Garg, S. K. and R. Buyya Green cloud computing and environmental sustainability. In S. Murugesan and G. R. Gangadharan (eds.), *Harnessing Green IT: Principles and Practices*. 2012, pp. 315–340.
10. Kliazovich, D., P. Bouvry, and S. U. Khan GreenCloud: A packet-level simulator of energy-aware cloud computing data centers. *The Journal of Supercomputing* 62(3): 1263–1283, 2012.
11. Baikie, B. and L. Hosman Green cloud computing in developing regions Moving data and processing closer to the end user. *Telecom World (ITU WT), 2011 Technical Symposium at ITU*, Geneva, Switzerland, October 24–27, 2011, pp. 24–28.

Further Reading

Gupta, P. and S. Gupta Mobile cloud computing: The future of cloud. *International Journal of Advanced Research in Electrical, Electronics and Instrumentation Engineering (IJAREEIE)* 1(3), September 2012.

Fernando, N., S. W. Loke, and W. Rahayu Mobile cloud computing: A survey. *Future Generation Computer Systems* 29(1): 84–106, 2013.

Cloud Incubator While Paper – DMTF, A White Paper from the Open Cloud Standards Incubator.

http://us.cdn4.123rf.com/168nwm/.

Index

A

Additive increase/multiplicative
    decrease (AIMD), 257
ADE, *see* Application development
    environment (ADE)
Agile software process
  advantages, 131
  cloud computing and virtualization,
    132–133
  continuous improvement, 131
  Salesforce.com's R&D organization,
    133–134
  SDLC, 129–130
  software projects, 130–131
Amazon Elastic Compute Cloud
    (Amazon EC2)
  computing service, 280–281
  ELB and EBS, 282–283
  and GoGrid Cloud, 307
  instance types, 281–282
  Manjrasoft Aneka, 154
  virtualization, 254–255
Amazon Simple Queue Service
    (Amazon SQS), 280, 283–284
Amazon Simple Storage Service
    (Amazon S3), 280, 283
Amazon Web Services (AWS)
  Amazon EC2, 280–281
  Amazon S3, 283
  Amazon SQS, 283–284
  data centers, 281
  Eucalyptus, 303
  IaaS providers, 76
  remote computing services, 280
  SLA violation, 30
Aneka
  description, 294
  features, 294–295
  Manjrasoft, 154
  map reduce model, 203
  platform, 294–295
  task execution model, 202–203
  thread execution model, 203

Apache CloudStack, 76, 305–306
Apache VCL
  architecture, 310–311
  compute nodes, 311
  conceptual overview, 311–312
  database server, 310
  description, 310
  management node, 311
  web server, 310
App Engine, Google
  costs, 280
  description, 279–280
  Java runtime environment, 279
  modules, 279–280
  PaaS providers, 17, 84
  web based applications, 153
APIs, *see* Application programming
    interfaces (APIs)
Application development environment
    (ADE)
  advantages, 152
  agile development, 150
  APIs, 154
  building software applications, 149
  cloud computing, 150–151
  desktop development, 151–152
  distributed development, 150
  Force.com, 153
  IBM, 154
  Intel, 154
  Manjrasoft Aneka, 154
  Rackspace, 154
  standards based, 149
  Windows Azure, 153, 284–285
Application programming interfaces
    (APIs)
  CloudI, 198–199
  Cloud Server, 292
  GoGrid, 307
  IaaS services, 69
  IBM, 154
  integration, 87
  Intel, 154
  PaaS providers, 83–85